Social Choice and the Mathematics of Manipulation

Honesty in voting, it turns out, is not always the best policy. Indeed, in the early 1970s, Allan Gibbard and Mark Satterthwaite, building on the seminal work of Nobel Laureate Kenneth Arrow, proved that with three or more alternatives there is no reasonable voting system that is non-manipulable; voters will always have an opportunity to benefit by submitting a disingenuous ballot. The ensuing decades produced a number of theorems of striking mathematical naturality that dealt with the manipulability of voting systems. This book presents many of these results from the last quarter of the twentieth century – especially the contributions of economists and philosophers – from a mathematical point of view, with many new proofs. The presentation is almost completely self-contained and requires no prerequisites except a willingness to follow rigorous mathematical arguments.

ALAN D. TAYLOR is the Marie Louise Bailey Professor of Mathematics at Union College, where he has been since receiving his Ph.D. from Dartmouth College in 1975. His research interests have included logic and set theory, finite and infinitary combinatorics, simple games, and social choice theory. He is the author of *Mathematics and Politics: Strategy, Voting, Power, and Proof* and coauthor of *Fair Division: From Cake-Cutting to Dispute Resolution* and *The Win-Win Solution: Guaranteeing Fair Shares to Everybody* (both with Steven J. Brams) and *Simple Games: Desirability Relations, Trading, and Pseudoweightings* (with William S. Zwicker).

OUTLOOKS

PUBLISHED BY CAMBRIDGE UNIVERSITY PRESS AND
THE MATHEMATICAL ASSOCIATION OF AMERICA

Mathematical content is not confined to mathematics. Eugene Wigner noted the unreasonable effectiveness of mathematics in the physical sciences. Deep mathematical structures also exist in areas as diverse as genetics and art, finance and music. The discovery of these mathematical structures has in turn inspired new questions within pure mathematics.

In the *Outlooks* series, the interplay between mathematics and other disciplines is explored. Authors reveal mathematical content, limitations, and new questions arising from this interplay, providing a provocative and novel view for mathematicians, and for others an advertisement for the mathematical outlook.

Social Choice and the Mathematics of Manipulation

ALAN D. TAYLOR
Union College

CAMBRIDGE UNIVERSITY PRESS
Cambridge, New York, Melbourne, Madrid, Cape Town, Singapore, São Paulo

Cambridge University Press
40 West 20th Street, New York, NY 10011-4211, USA

www.cambridge.org
Information on this title: www.cambridge.org/9780521810524

MATHEMATICAL ASSOCIATION OF AMERICA
1529 Eighteenth St. NW, Washington, DC 20036, USA

First published 2005

Printed in the United States of America

A catalog record for this publication is available from the British Library.

Library of Congress Cataloging in Publication Data

Taylor, Alan D., 1947–

Social choice and the mathematics of manipulation / Alan D. Taylor.
p. cm. – (Outlooks)

Includes bibliographical references and index.

ISBN 0-521-81052-3 (hardback : alk. paper) – ISBN 0-521-00883-2 (pbk. : alk. paper)

1. Voting – Mathematical models. 2. Social choice – Mathematical models.
3. Political Sciences – Mathematical models. 4. Game theory.
I. Title. II. Series.

JF1001. T39 2005
324.6′01′5193–dc22 2004054521

ISBN-13 978-0-521-81052-4 hardback
ISBN-10 0-521-81052-3 hardback

ISBN-13 978-0-521-00883-9 paperback
ISBN-10 0-521-00883-2 paperback

Contents

Preface

Were honesty always the best policy, this indeed might be a better world. But there seems to be a place for the little white lie, and there is certainly reason for many supporters of Ralph Nader in the state of Florida – as they watched Albert Gore concede the U.S. presidential election to George W. Bush on the evening of December 12, 2000 – to regret having cast sincere ballots, the result of which was a victory for their third choice (Bush) instead of their second choice (Gore).

We have nothing to say here about the little white lie. In this book, however, we collect much of what is known regarding a single fundamental question of obvious political importance and surprising mathematical naturality: In what election-theoretic contexts is honesty in voting the best policy?

For example, consider an election in which there are three or more candidates from which a unique winner must be chosen, and in which each voter casts a ballot that gives his or her ranking of the candidates from best to worst with no ties. Can one, in this situation, devise a voting procedure such that each candidate wins at least one hypothetical election and with which no voter can ever gain by unilaterally changing his or her ballot?

As stated, this turns out to be a trivial question. If there are n voters, then a moment's reflection reveals n such voting procedures, each obtained by fixing one of the voters and taking the winner to be his or her top-ranked candidate. Dismissing these – they are, after all, dictatorships – leaves the better question: Are there any others?

The answer, quite remarkably, is no. This is precisely what the Gibbard–Satterthwaite theorem of the early 1970s asserts, and it is this result that gives rise to most of what follows in this book. That theorem is related to – indeed, some would say equivalent to – the celebrated 1950 result known as Arrow's impossibility theorem.

If there is a weakness to the Gibbard–Satterthwaite theorem, it is the assumption that winners are unique. But if we drop the uniqueness of winners as an assumption, then there are voting systems that intuitively seem to be non-manipulable. For example, the voting procedure that declares everyone to be tied for the win regardless of the ballots (a very uninteresting example) or the one that takes as winners all candidates with at least one first-place vote (certainly a more interesting example).

Why do we speak of these two procedures as being only "intuitively" non-manipulable? The problem is that if a voter's preferences are given by a list, then it is not at all clear what it means to say that he or she prefers one *set* of candidates to another *set* of candidates. For example, if a voter ranks alternative a over alternative b over alternative c over alternative d, does he or she then prefer the set $\{a, d\}$ to $\{b, c\}$ or vice versa? It's certainly not obvious.

Thus, one of our objectives is to collect many of the definitions, theorems, and questions that arise when one asks about single-voter manipulability in election-theoretic contexts in which winners are not necessarily unique. Most of the results we present – whether in the concrete setting of the twenty voting rules that we introduce in Chapter 1 or the more abstract context of theorems like that of Gibbard and Satterthwaite – are organized around four kinds of manipulability that we call, from strongest to weakest, single-winner manipulability, weak-dominance manipulability, manipulability by optimists and pessimists, and expected-utility manipulation.

Our undertaking is interdisciplinary in the sense that it is, in large part, a mathematician's presentation of some major contributions that economists and philosophers have made to the field of political science. Thus, few of the results in this book originated with the author, but many of the proofs did. For example, we give unified proofs of three important manipulability results in three of the major voting-theoretic contexts: the Gibbard–Satterthwaite theorem in the context wherein the outcome of an election is a single winner (Chapter 3), the Duggan–Schwartz theorem in the context wherein the outcome of an election is a set of winners (Chapter 4), and the Barberá–Kelly theorem in the context wherein the outcome of an election is a choice function (Chapter 5).

There are virtually no prerequisites for reading this book, except that a certain degree of what is usually called mathematical maturity is required beginning with Chapter 3. The nine chapters in the book are organized into three parts, each consisting of three chapters. We comment on each part in turn.

Part I of the book is presented at a suitably accessible level for use in a number of undergraduate or graduate courses in mathematics, economics, and political science. In particular, Chapter 1 is an introduction to social choice theory that provides (i) an explicit discussion of the different contexts in which one works; (ii) something of the history of the field; (iii) accurate statements of Arrow's impossibility theorem for voting rules, social choice functions, and social welfare functions; and (iv) a wide range of examples of voting rules. Chapter 2 is an introduction to the notion of manipulability, largely in the context of twenty specific voting procedures, and Chapter 3 presents a careful proof of the Gibbard–Satterthwaite theorem and deals with the question of the extent to which it is equivalent to Arrow's theorem.

There are more than seventy-five exercises in Part I, and these range from routine verifications to additional development of the material in the chapter, with hints (or outlines) provided where needed. Each exercise is labeled with a "C" for computational, an "S" for short answer, or a "T" for theory. Hopefully these labels will be of some use, but all three terms are being used somewhat metaphorically.

Part II of the book begins with a treatment of manipulation in the contexts wherein the outcome of an election is a set of winners (Chapter 4) and a social choice function (Chapter 5). Each chapter contains a direct proof of the main result that mimics what was done with the Gibbard–Satterthwaite theorem in Chapter 3 by combining a new idea – down-monotonicity for singleton winners – with a number of classical ideas, most of which arose in later treatments of Arrow's impossibility theorem.

In Chapter 6, we move to the case of infinitely many voters, and we present the known ultrafilter versions of Arrow's theorem and the Gibbard–Satterthwaite theorem, the latter of which requires finitely many alternatives and coalitional non-manipulability. We also provide an extension to the case where there are infinitely many alternatives as well as infinitely many voters. The proofs in Chapter 6 are completely self-contained, though less pedestrian than the presentations in Part I.

Part III of the book contains a number of additional results in cases of both single winners (Chapter 7) and multiple winners (Chapter 8). Finally, in Chapter 9, we conclude the book with a brief treatment of some voting-theoretic situations in which ballots and election outcomes are different from those of Chapters 1–8.

PART ONE

1

An Introduction to Social Choice Theory

1.1 Some Intuitions, Terminology, and an Example

In a capitalist democracy there are, according to Nobel Laureate Kenneth J. Arrow (Arrow, 1950), "essentially two ways by which social choices can be made: voting, typically used to make 'political' decisions, and the market mechanism, typically used to make 'economic' decisions." Our concern here is exclusively with the former.

Thus, for us, democratic theory is, in the words of Peter C. Fishburn (Fishburn, 1973, p. 3), "based on the premise that the resolution of a matter of social policy, group choice, or collective action should be based on the preferences of the individuals in the society, group, or collective." And social choice theory is, as William H. Riker put it (Riker, 1986, p. xi), "the description and analysis of the way that the preferences of individual members of a group are amalgamated into a decision of the group as a whole." Arrow, by the way, is an economist, Fishburn a mathematician, and Riker a political scientist.

Let's start with a very simple example. Suppose we have an academic department with ten faculty members, one of whom is serving as chair. They are in the process of filling a position in the department and have interviewed five finalists for the job. Needless to say, the different department members disagree on the ranking of the five, and what is needed is some procedure for passing from the preferences of the individuals in the department to the "preferences," if you will, of the group.

First, let's ask what the ballots could look like. If we were to opt for simplicity, a ballot would have just a single name on it, representing (we presume, perhaps naïvely) the candidate who is that department member's top choice. Or, we could allow a ballot to contain several names, intuitively representing either a group that this department member feels is tied for the top, or those candidates that the department member finds acceptable ("approves of" in the parlance of the voting system called approval voting, which we discuss later).

But none of these ballot types is very expressive, and, in voting situations such as we are describing, one tends to use ballots that allow each department member to rank-order the candidates from best to worst, in his or her opinion, perhaps allowing ties (representing indifference) in the individual ballots and perhaps not.[1] Intuitively, this description of a ballot is fine, but some of what we do in this book requires a bit more precision. So let us momentarily set this example to one side and introduce some notation and terminology.

We make use of the universal quantifier "∀" meaning "for all" and the existential quantifier "∃" meaning "there exists." We do not, however, display or abbreviate the phrase "such that," although it is almost always required in reading an expression such as

$$\forall x \in \mathrm{A}\, \exists y \in \mathrm{A}\; x\mathrm{P}y.$$

We also use the standard abbreviation "iff" for "if and only if."

Set-theoretically, $|\mathrm{A}|$ denotes the number of elements in the finite set A, and $\wp(\mathrm{A})$ is collection of all subsets of A. If n is a positive integer, then $[\mathrm{A}]^n = \{\mathrm{X} \in \wp(\mathrm{A}): |\mathrm{A}| = n\}$. Any subset R of $\mathrm{A} \times \mathrm{A}$ is a *binary relation* on A, and in this case we write "$a\mathrm{R}b$" or we say "$a\mathrm{R}b$ holds" to indicate that $(a, b) \in \mathrm{R}$, and we write "$\neg(a\mathrm{R}b)$" or say "$a\mathrm{R}b$ fails" to indicate that $(a, b) \notin \mathrm{R}$. Finally, if R is a binary relation on A and $v \subseteq \mathrm{A}$, then the *restriction of R to v*, denoted $\mathrm{R}|v$, is the binary relation on v given by $\mathrm{R}|v = \mathrm{R} \cap (v \times v)$.

The binary relations we are most concerned with satisfy one or more of the following properties.

[1] Most readers will assume that the picture we are painting is one in which the notion of indifference is transitive, and we will, in fact, be adopting that convention. Fishburn (1973, pp. 5 and 6), on the other hand, spends considerable time with the case in which indifference is intransitive, and justifiably so. As an example of intransitive indifference in our present context, suppose that Applicant B might receive (and bring along) a large research grant for which he or she has applied, and that we in the department will not know whether or not this grant application is successful before the job offer will be made. We could handle this by pretending to have six applicants instead of five, with Applicant B split into "Applicant B without research support" and "Applicant B with research support." It is now easy to imagine a situation in which a department member might be indifferent between Applicant C and either of these choices. But of course, anyone would (presumably) prefer Applicant B with research support to Applicant B without research support.

Definition 1.1.1. A binary relation R on a set A is:

reflexive	if	$\forall x \in A$, $x R x$
irreflexive	if	$\forall x \in A$, $\neg(x R x)$
symmetric	if	$\forall x, y \in A$, if $x R y$, then $y R x$
asymmetric	if	$\forall x, y \in A$, if $x R y$, then $\neg(y R x)$
antisymmetric	if	$\forall x, y \in A$, if $x R y$ and $y R x$, then $x = y$
transitive	if	$\forall x, y, z \in A$, if $x R y$ and $y R z$, then $x R z$
complete	if	$\forall x, y \in A$, either $x R y$ or $y R x$

Definition 1.1.2. A binary relation R on a set A is a *weak ordering (of A)* if it is transitive and complete and a *linear ordering (of A)* if it is also antisymmetric.

If R is a weak ordering of A, then the completeness of R implies (letting $x = y$) that R is also reflexive. Intuitively, a weak ordering corresponds to a list with ties, with $x R y$ being thought of as asserting that x is at least as good as y. A linear ordering corresponds to a list without ties, with $x R y$ now being thought of as asserting that either $x = y$ or x is strictly better than y.

Associated to each weak ordering R of A, there are two so-called derived relations P and I.

Definition 1.1.3. If R is a weak ordering of A, then the derived relations of *strict preference* P and *indifference* I are arrived at by asserting that $x P y$ iff $\neg(y R x)$ and $x I y$ iff $x R y$ and $y R x$.

If R is a linear ordering of A, then the derived relation I is just equality, and the derived relation P is referred to as a *strict linear ordering of A*. Exercise 1 asks for verification that if R is a weak ordering, then the derived relation P of strict preference is transitive and asymmetric (and thus irreflexive), while the derived relation I of indifference is reflexive, symmetric, and transitive. Thus, I is an equivalence relation, and P is a strict linear ordering of the I-equivalence classes.

The following definition uses the concepts of weak and linear orderings to formalize some election-theoretic terminology.

Definition 1.1.4. If A is a finite non-empty set (which we think of as the set of alternatives from which the voters are choosing), then an *A-ballot* is a weak ordering of A. If, additionally, n is a positive integer (where we think of $N = \{1, \ldots, n\}$ as being the set of voters), then an *(A, n)-profile* is an n-tuple of A-ballots. Similarly, a *linear A-ballot* is a linear ordering of A, and a *linear (A, n)-profile* is an n-tuple of linear A-ballots.

When the set A of alternatives and the integer n are clear from the context, we will use "ballot" in place of "A-ballot" and "profile" in place of "(A, n)-profile." Similarly, we will often use a phrase like "If **P** is a profile" as an abbreviation for the phrase "If **P** is an (A, n)-profile for some set A and some positive integer n." This latter remark is illustrated in the next paragraph.

If **P** is a profile, then R_i denotes its ith component (that is, the ballot of the ith voter), with P_i and I_i denoting the corresponding derived relations of strict preference and indifference for the ith voter. When we have several profiles under consideration at the same time, we use names such as **P**, **P**$'$, and **P**$''$, with the understanding that their components and the derived relations also carry the prime, double prime, etc.

If $\mathbf{P} = <R_1, \ldots, R_n>$ is an (A, n)-profile and $X \subseteq A$, then the restriction of **P** to X, denoted **P**|X, is the profile $<R_1|X, \ldots, R_n|X>$. If $i \in N$, then $\mathbf{P}|N - \{i\}$ is the profile $<R_1, \ldots, R_{i-1}, R_{i+1}, \ldots, R_n>$. This dual use of the vertical bar should cause no confusion.

The following definition collects some additional ballot-theoretic notation we will need.

Definition 1.1.5. Suppose **P** is a linear (A, n)-profile, X is a set of alternatives (that is, $X \subseteq A$), and i is a voter (that is, $i \in N$). Then:

$\text{top}_i\,(\mathbf{P}) = x$	iff	$\forall y \in A\ xR_iy$
$\max_i\,(X, \mathbf{P}) = x$	iff	$x \in X$ and $\forall y \in X\ xR_iy$
$\min_i\,(X, \mathbf{P}) = x$	iff	$x \in X$ and $\forall y \in X\ yR_ix$

Thus, $\max_i(X, \mathbf{P})$ is the element of X that voter i has most highly ranked on his or her ballot in **P**, and $\min_i(X, \mathbf{P})$ is the one ranked lowest. $\text{Top}_i(\mathbf{P})$ is the alternative that voter i has at the top of his or her ballot in **P**. Hence, $\text{top}_i(\mathbf{P}) = \max_i(A, \mathbf{P})$ and $\max_i(X, \mathbf{P}) = \text{top}_i(\mathbf{P}|X)$.

Once we have decided what the ballots will look like, it might well seem natural to ask what we *do* with these ballots to find a winner. That, however, is somewhat getting ahead of ourselves. What we really need to decide first is what kind of outcome our balloting should yield.

For example, should the outcome in the departmental hiring example that we considered earlier be a single winner with the understanding that the chair will call that candidate with the offer, and if he or she refuses, then the chair will reconvene the department and start the balloting process all over again? Or do we allow ties in the outcome of the balloting, with the understanding, perhaps, that either the chair or the dean will be allowed to break the tie?

From the chair's point of view, perhaps the most desirable outcome is neither of these, but instead a list without ties – a linear ordering – that gives the department's "overall ranking" (an intuitive phrase, at best) of the candidates. The chair can then begin at the top, and make the calls one by one until the position is accepted. Similarly, if the outcome is a list with ties – a weak ordering – then we could once again agree to give either the chair or the dean tie-breaking power.

But suppose we are in a situation wherein we proceed to individually rank-order all the candidates, and then, while the chair is handling the accompanying administrative tasks involving the dean and the college affirmative action officer, several candidates notify the department that they have accepted other offers and no longer wish to be considered. To handle such contingencies, we might want the outcome of the election to be a "choice function" that selects one or more "winners" from each non-empty subset of candidates.

This last possibility is an interesting one, and later in this chapter we say something about the historical perspective from the field of economics that apparently played a role in the prominence of so-called choice functions in the formalism. But for now, we conclude the present section with precise definitions of the different kinds of "social choice procedures" alluded to earlier.[2]

Definition 1.1.6. Suppose that A is a non-empty set, n is a positive integer, and V is a function whose domain is the collection of all (A, n)-profiles.[3] Then V is:

(1) a *resolute voting rule for* (*A*, n) if, for every (A, n)-profile **P**, the election outcome V(**P**) is a single element of A,

[2] There is, unfortunately, no common terminology in the literature for the concepts in Definition 1.1.6. Our use of "voting rule" and "social welfare function" is quite common and used, for example, in Moulin (2003). Our use of "social choice function" for a voting rule with a variable agenda is also not without precedent, but one also sees variants of this with, for example, "decision" used in place of "choice" and/or "procedure" used in place of "function." Our use of "resolute" to mean "without ties" would not, however, be considered standard, although it has appeared in the work of Duggan and Schwartz (1993, 2000) and goes back at least to Gärdenfors (1976).

[3] We are building into our formalism a condition known in the literature as "unrestricted scope." Intuitively, this asserts that no voter should be prohibited from submitting any ballot. There is something to be gained (e.g., conceptual simplicity) by burying certain assumptions within the formalism, but there is also a loss. For example, our choice to build in unrestricted scope leads to an omission of a number of important results related to the following question: What conditions can one impose on a profile that will ensure that "bad things" (e.g., opportunities for manipulation or cycles wherein a majority of voters prefer a to b, a majority prefers b to c, and yet a majority also prefers c to a) don't happen? One answer to this, by the way, involves "single-peaked preferences" – see Black (1958), Sen (1966), Fishburn (1973), Taylor (1995), and Shepsle and Bonchek (1997).

(2) a *voting rule for* (*A,* n) if, for every (A, n)-profile **P**, the election outcome V(**P**) is a non-empty[4] subset of A,
(3) a *social choice function for* (*A,* n) if, for every (A, n)-profile **P**, the election outcome V(**P**) is a "choice function" C that picks out a non-empty subset C(v) of v for each non-empty subset v of A,
(4) a *resolute social choice function for* (*A*, n) if, for every (A, n)-profile **P**, the election outcome V(**P**) is a choice function C that picks out a single element C(v) from v for each non-empty subset v of A,
(5) a *social welfare function for* (*A*, n) if, for every (A, n)-profile **P**, the election outcome V(**P**) is a weak ordering of A, and
(6) a *resolute social welfare function for* (A, n) if, for every (A, n)-profile **P**, the election outcome V(**P**) is a linear ordering of A.

The set v of alternatives occurring in (3) and (4) is called an *agenda*.[5] In general, V is called an *aggregation procedure* if it is any one of (1)–(6). As before, we suppress the reference to the pair (A, n) whenever possible.

In point of fact, our primary concern is with the first three aggregation procedures given in Definition 1.1.6:

(1) resolute voting rules (Chapter 3 and 7)
(2) (non-resolute) voting rules (Chapters 4 and 8)
(3) social choice functions (Chapters 5 and 8).

With each kind of aggregation procedure, there are two contexts in which we work: the one in which only linear ballots are considered and the other in which we allow ties in the ballots.

But before we press on with any additional notation and terminology, let us pause to give a quick historical overview of the field of social choice theory. This will, at the same time, provide an informal introduction to a number of aggregation procedures, most of which are rigorously defined in Section 1.4.

[4] By saying "non-empty" we are disallowing the possibility of an election resulting in no alternative being chosen. Fishburn (1973, p. 3) justifies this by the observation that one can always include alternatives such as "delay the decision to a later time" or "maintain the status quo."

[5] There are at least three different ways the term "agenda" is used in voting-theoretic contexts: (1) as the set of alternatives from which a choice is to be made (our use here), (2) as an ordering in which alternatives will be pitted against each other in one-one-one contests based on the ballots cast, and (3) as an ordering in which alternatives will be pitted against the status quo until one defeats the status quo.

1.2 A Little History

Jean Charles Chevalier de Borda (1733–99) was, according to Duncan Black (1958, p. 156), "the first thinker to develop a mathematical theory of elections." In Borda's 1781 paper (apparently the only one of his that we now possess) he introduced the aggregation procedure that is known today as the *Borda count*. It selects a winner (or winners) from among k alternatives by assigning each alternative k-1 points for each ballot on which it appears first, k-2 points for each ballot on which it occurs second, and so on. The points are then summed, with the winner (or winners) being the alternative with the most points. If ties in the ballots are allowed, the procedure can be suitably modified. These so-called Borda scores can also be used to produce a list, perhaps with ties, as the outcome of an election. Interestingly, recent historical work by McLean and Urken (1993) and Pukelsheim (unpublished) reveals that Borda's system had been explicitly described in 1433 by Nicholas of Cusa (1401–64), a Renaissance scholar interested in the question of how German kings should be elected.

In that same 1781 paper, Borda pointed out a very nice equivalent version of the Borda count, not often referred to today, that goes as follows: Each alternative is pitted one-on-one against each of the other alternatives, based on the ballots cast. Having done this, one doesn't just look for the alternative that defeats the most other alternatives – this would be quite a different social choice procedure, one known today as *Copeland's function* and introduced in an unpublished 1951 note by A. H. Copeland (Fishburn, 1973, p. 170). Instead, one looks for the alternative with the greatest total score from these one-on-one contests. For example, if one of four alternatives defeats two others by scores of 4–3 and 5–2, but loses to the third by a score of 6–1, then that alternative's total score is $4 + 5 + 1 = 10$.

In fact, this latter characterization of the Borda count gives rise to an easy way to hand-calculate Borda scores given a sequence of ballots: Given an alternative a, one simply counts the total number of occurrences of other alternatives below a, proceeding ballot-by-ballot (Taylor, 1995). It is easy to see that this is the same as Borda's equivalent, the difference being that what we are suggesting here is a ballot-by-ballot enumeration instead of an alternative-by-alternative enumeration.

But Borda was not alone in his election-theoretic ponderings, as a systematic theory of elections was, as Black (1958, p. 156) again informs us, "part of the general uprush of thought in France in the second half of the eighteenth century." For example, Borda's "method of marks" arose again in 1795 in the writings of Pierre-Simon, Marquis de Laplace (1749–1827), who derived the method via some probabilistic considerations.

However, no one's contributions at that time were more important than the observations of Marie Jean Antoine Nicolas Caritat, Marquis de Condorcet (1743–94). In a 1785 publication (Condorcet, 1785) he explicitly discussed what we now call the "Condorcet voting paradox" wherein we find that if a group of voters is broken into three equal-size groups with preferences for three alternatives as shown below, then a majority prefers a to b, a majority prefers b to c, but (somewhat paradoxically) a majority also prefers c to a.

Group #1	Group #2	Group #3
a	b	c
b	c	a
c	a	b

The Condorcet Voting Paradox

For a given sequence of ballots, a candidate that would defeat each of the other candidates in a one-on-one contest – based on the ballots – is called a *Condorcet winner* for that election. For example, if we regard the U.S. presidential race of 2000 in the state of Florida as one with four candidates (Bush, Gore, Nader, and Buchanan), then it is almost certainly true that Gore was the Condorcet winner. Of course, Bush won using what is known as plurality voting, wherein one simply looks for the alternative with the most first-place votes.

The name "Condorcet's method" is often applied to the voting procedure in which the Condorcet winner, if there is one, is the unique winner of the election and there is no winner otherwise. Like Borda, Condorcet was anticipated by several centuries. Indeed, very recent research by Pukelsheim et al. (unpublished) shows that Condorcet's method can be traced back at least to Ramon Llull (1232–1316), a Catalan philosopher and missionary who was involved in devising election schemes for selecting the abbess of a convent. (See Pukelsheim's amusing "Spotlight" on page 418 of COMAP, 2003.)

Condorcet knew of Borda's work and Condorcet pointed out in his writings that Borda's method of marks, like plurality voting, can result in a Condorcet winner not being elected. Nevertheless, even Condorcet would probably have been surprised that this "defect" would one day determine a presidential election, as it did in the United States in the year 2000.

The nineteenth century saw a few small election-theoretic results from people like Issac Todhunter (1820–84), M. W. Crofton (1826–1915), E. J. Nanson (1850–1936), and Francis Galton (1822–1911). But it was the Reverend Charles Lutwidge Dodgson (1832–98) – better known by the pseudonym Lewis Carroll – who made the most significant contributions at the time, beginning with his rediscovery in 1874 of the Condorcet voting paradox. Dodgson was the

Mathematical Lecturer at Christ Church, and he even published a monograph entitled *Elementary Treatise on Determinants* between the appearance of *Alice's Adventures in Wonderland* (1865) and *Through the Looking Glass, and What Alice Found There* (1872). The mathematical biographer E. T. Bell spoke of Dodgson as having in him "the stuff of a great mathematical logician" (Black, 1958, p. 195), and Duncan Black characterized Dodgson's understanding of the theory of elections and committees as "second only to that of Condorcet" (Black, 1958, p. 212).

In an 1873 pamphlet entitled "A Discussion of the Various Methods of Procedure in Conducting Elections" (Black, 1958, p. 214), Dodgson proposes – without claiming to have discovered them himself – several "Methods of Procedure" for the case where an election is necessary. The description of each that follows is taken verbatim from that pamphlet, although we do not reproduce his examples showing why he finds fault with each. Our own comments are added in brackets.

(1) The Method of a Simple Majority: In this Method, each elector names the one candidate he prefers, and he who gets the greatest number of votes is taken as the winner. [This is known today as *plurality voting*.]
(2) The Method of an Absolute Majority: In this Method, each elector names the one candidate he prefers; and if there be an absolute majority for any one candidate, he is the winner. [Dodgson offers no provision for the case where no one has more than half the votes.]
(3) The Method of Elimination, where the names are voted on by two at a time: In this Method, two names are chosen at random and proposed for voting, the loser is struck out from further competition, and the winner taken along with some other candidate, and so on, til there is only one candidate left. [This procedure is essentially what Straffin (1980) calls "sequential pairwise voting with a fixed agenda" (see also Taylor, 1995). Here, "agenda" refers to an ordering of the alternatives.]
(4) The Method of Elimination, where the names are voted on all at once: In this Method, each elector names the one candidate he prefers: the one who gets the fewest votes is excluded from further competition, and the process is repeated. [This is the procedure introduced in 1861 by Thomas Hare, and known today by various names including the "Hare system" and the "single transferable vote system." In 1862, John Stuart Mill (Mill, 1862) spoke of it as being "among the greatest improvements yet made in the theory and practice of government." It is currently used to elect public officials in Australia, Malta, the Republic of Ireland, and Northern Ireland. The Hare system was essentially the method used to choose Sydney, Australia, as the site of the 2000 Summer Olympics. In this election, Beijing would have been the plurality winner, but after Istanbul, Berlin, and Manchester were eliminated (in that order), Sydney defeated Beijing by a vote of 45 to 43.]
(5) The Method of Marks: In this Method, a certain number of marks is fixed, which each elector shall have at his disposal; he may assign them all to one candidate,

> or divide them among several candidates, in proportion to their eligibility; and the candidate who gets the greatest total number of marks is the winner. [This procedure is certainly not the same as Borda's "method of marks," but, if we ignore the phrase "in proportion to their eligibility," it is a method known today as "cumulative voting" and championed by people such as Lani Guinier (nominated by President Clinton in 1993 to be the administration's civil rights enforcement officer) as one way to give minorities more power. Dodgson later proposed a modification of this procedure that does, in fact, coincide with the Borda count.]

In later writings, Dodgson advocated the choice of a Condorcet winner if one existed, but he was unsure of the best course of action if there were cyclical preferences as in the Condorcet voting paradox. Remarkably, there seems to be little new in terms of concrete voting procedures forthcoming in the 100 years following Dodgson's 1873 pamphlet, with the emergence of "approval voting" in the late 1970s as perhaps the next procedure proposed that received somewhat wide attention (Brams and Fishburn, 1983).

Academics, however, have proposed other systems that are radically different from what we usually think of (plurality voting, the Borda count, the Hare system, etc.) and, although perhaps of more theoretical interest than practical interest, they are certainly not without their charms. We'll mention two, but the reader is warned that our descriptions are intentionally terse – the idea being only to convey that there exist non-trivial procedures that any general result must encompass.

The first example arises in some work of Hervé Moulin (Moulin, 1981 or 1983, p. 407), and we give here only a simplified version. Suppose that we have n voters and n alternatives. Given a profile **P**, consider the voting system in which an alternative a fails to be among the winners iff there exists a set X of voters and a set B of alternatives such that $|\mathrm{X}| + |\mathrm{B}| > n$, and every voter in X ranks every alternative in B higher on his or her ballot than a. The intuition here (roughly) is to give sets of voters the power to veto sets of alternatives if the set of alternatives is proportionately smaller than the set of voters and to reject any alternative that, if chosen, would trigger the use of such veto power. A non-trivial theorem of Moulin asserts (in part) that there is always at least one winner for each profile. For more on this, see Exercises 3 and 4.

The second example is due to Dan Felsenthal and Moshé Machover (Felsenthal and Machover, 1992). Given a profile **P** for the set $\mathrm{A} = \{a_1, \ldots, a_k\}$ of alternatives, consider the $k \times k$ matrix G whose (i, j)th entry is $+1$ if a_i defeats a_j in a one-on-one contest using these ballots, is -1 if a_i loses to a_j in a one-on-one contest using these ballots, and is 0 otherwise. Then G determines a two-person game between Row and Column, with a strategy for either player being a probabilistic rule for selecting a candidate, and thus represented by a row vector $(p_1, \ldots, p_k)$, where p_i is the probability of choosing candidate a_i. Standard game-theoretic results (e.g., using linear programming techniques) now yield the existence of an optimal strategy $(p_1, \ldots, p_k)$. The winner of the election is then determined by a weighted lottery in which candidate a_i has probability p_i of being chosen.

The Felsenthal–Machover procedure does not correspond to a voting rule according to Definition 1.1.6 because it involves a weighted lottery to select the winner and thus is not a function. We could, however, pick a value $q \in [0, 1/k]$ and say that a_i is among the winners if, in the optimal strategy, $p_i \geq q$.

By the way, a simple version of the Felsenthal–Machover procedure immediately suggests itself for use in, say, presidential elections in the United States. We could simply end each election with a lottery in which the weights are determined by the popular vote. Of course, one would not want the correspondence to be linear. For example, if one candidate received 60% of the popular vote, his or her weighting in the lottery might well be .99. One benefit of this system is that it would make elections less "chaotic," both literally (a small change in initial conditions – the ballots – would give rise to only a small change in election outcome – the probabilities used in the lottery) and figuratively (there would have been no recounts or court cases in the U.S. presidential election of 2000, because it would clearly have been a 50–50 lottery between Bush and Gore). The world, however, is certainly not ready for any proposal as radical as this one.

We'll have a great deal more to say about specific voting procedures in the last section of this chapter. But let us now turn to what is generally regarded as the seminal result in the field of social choice.

1.3 Arrow's Theorem

For an area of study to become a recognized field, or even a recognized subfield, two things are required: It must be seen to have coherence, and it must be seen to have depth. The former often comes gradually, but the latter can arise in a single flash of brilliance. Game theory, for example, came into existence, in part, because of the depth provided by John von Neumann's proof of the minimax theorem (von Neumann, 1928) and the coherence provided by *Games and Economic Behavior* (von Neumann and Morgenstern, 1944).

With social choice theory, there is little doubt as to the seminal result that made it a recognized field of study: Arrow's impossibility theorem. Kenneth J. Arrow is the Joan Kenney Professor of Economics and Professor of Operations Research, Emeritus at Stanford University and the 1972 recipient of the Nobel Memorial Prize in Economic Science.

The fundamental result on manipulability – the Gibbard–Satterthwaite theorem – is related to Arrow's theorem in important ways. This more than justifies a consideration of the latter here, but such a discussion will also give us an opportunity to illustrate the different formalizations of a social choice procedure given in Definition 1.1.6.

Everyone has his or her own take on Arrow's impossibility theorem. There is commonality for sure – everyone sees it as asserting the impossibility of finding a function that will amalgamate a sequence of individual ballots (represented by lists or lists with ties) in a "reasonable" way. Moreover, most agree that "reasonable" in this context involves three to five conditions, including the following two:

(1) Pareto (P): If every individual voter strictly prefers some particular alternative to another, then this unanimous preference is reflected in the outcome of the election (and we are purposefully leaving this statement somewhat vague for the moment).
(2) Nondictatorship (D): There is no voter who is a dictator in the sense that his or her strict preference is always reflected in the outcome of the election (again, purposefully left somewhat vague for the moment).

These two conditions do very little to limit the kind of amalgamation that springs to mind. The imprecision in our statements of P and D is necessitated by the fact that we have not yet specified whether we are talking about voting rules, social welfare functions, or social choice functions.

In fact, when authors such as Borda, Condorcet, and Dodgson give examples of different voting systems, it is often not clear within which of these contexts – voting rules, social welfare functions, or social choice functions – they are working. But this imprecision is not necessarily a bad thing. For example, the Borda count provides natural examples of the first two contexts (voting rules and social welfare functions) in the following ways:

(1) It yields a voting rule by simply taking the set of winners to be the alternatives in A that have the highest Borda score.
(2) It yields a social welfare function by using the Borda scores of the alternatives in A to obtain a weak ordering of A.

Interestingly, there are two natural ways that we can use the Borda count to get a social choice function on A. Amartya Sen (Sen, 1977 or 1982, p. 186) refers to these as "the narrow Borda rule" and the "broad Borda rule."

(3a) Given a set A, a profile **P**, and an agenda v, we can first restrict the ballots in **P** to v (erasing from the ballots all the alternatives that are not in v) and then calculate the Borda scores as usual, with the alternatives having the highest Borda scores winning.

(3b) Given a set A, a profile **P**, and an agenda v, we can first calculate the Borda scores using the ballots that list all the elements of A and then choose as winners the alternatives in v whose Borda scores are highest among the alternatives in v.

Notice that in both (3a) and (3b) we are doing two things: restricting to the set v and calculating Borda scores. The difference is in the order in which we do these. In (3a), we are first restricting (the ballots) to the set v and then calculating Borda scores; in (3b), we are first calculating Borda scores and then restricting (the set of winners) to the set v.

To see that both (3a) and (3b) are reasonable, let's return to our example wherein our academic department is trying to choose among five candidates they have interviewed. Suppose it is Friday morning and the department is meeting to decide whom the chair will call with a job offer that afternoon. Let's also assume that the department has agreed to use the Borda count (although we could repeat the argument to follow with the Hare system, plurality voting, etc.).

Now, the issue at hand is the following. Undoubtedly, as we have said, one or more of the five finalists is no longer available because he or she has decided to accept another offer and simply has not yet notified all of the other schools to which he or she applied. So, in fact, there is a subset v (hopefully non-empty) of the set A of five candidates consisting of those who would accept an offer from the chair if it were to be forthcoming.

Thus, the question becomes whether the chair first makes the five phone calls necessary to see which of the five candidates are still available and then brings the department together to use the Borda count (on that subset), or first brings the department together to use the Borda count on all five candidates in A, and then makes the phone calls that will reveal, if not all of v, at least the candidate in v ranked highest using the Borda count on all of A. Proceeding in either way seems quite reasonable, with the former way corresponding to (3a) and the latter to (3b).

In choosing a context in which to work, Arrow's starting point was to consider social choice functions. In part, this was a natural choice for Arrow because of his training as an economist. For example, he says in his famous 1950 paper (Arrow, 1950):

> We assume that there is a basic set of alternatives which could conceivably be presented to the chooser. In the theory of consumers' choice, each alternative would be a commodity bundle; in the theory of the firm, each alternative would be a

> complete decision on all inputs and outputs; in welfare economics, each alternative would be a distribution of commodities and labor requirements. These alternatives are mutually exclusive; they are denoted by small letters x, y, z, . . . On any given occasion the chooser has available to him a subset S of all possible alternatives, and he is required to choose one out of this set. The set S is a generalization of the well-known opportunity curve; thus, in the theory of consumer's choice under perfect competition, it would be the budget plane. It is assumed further that the choice is made in this way: Before knowing the set S, the chooser considers in turn all possible pairs of alternatives, say x and y, and for each pair he makes one and only one of three decisions: x is preferred to y, x is indifferent to y, or y is preferred to x. The decisions made for different pairs are assumed to be consistent with one another, so that, for example, if x is preferred to y and y to z, then x is preferred to z; similarly, if x is indifferent to y and y to z, then x is indifferent to z. Having this ordering of all possible alternatives, the chooser is now confronted with a particular opportunity set S. If there is one alternative in S which is preferred to all others in S, then the chooser selects that one alternative.

Thus, Arrow is working in a context in which the ballots are weak orderings. (Arrow is also starting with strict preference P and indifference I and later introducing the relation R by saying that xRy iff either xPy or xIy. The conditions he imposed on P and I guarantee that R is a weak ordering.) Moreover, the set S Arrow describes is playing the role of an agenda, so he seems to be working with a resolute social choice function. But Arrow later adds in a footnote that he recognizes that often there will be a set of several alternatives in S that are indifferent to each other but preferred to any other element of S. Hence, Arrow's context is really that of a (not necessarily resolute) social choice function.

Within this context, Arrow's impossibility theorem arises by adding two additional conditions to the two we previously had, which were Pareto (P) and non-dictatorship (D), with both P and D still requiring clarification that we give momentarily. The two conditions being added are:

Transitive Rationality (T): The social choice function is derived from a social welfare function in the sense that the winners are the elements of v that are ranked highest by the social welfare function.

Independence of Irrelevant Alternatives (IIA): The social choice function is such that the choice of winners from a subset v of A depends only on how the voters rank the alternatives in v. This is called "independence of infeasible alternatives" in Fishburn (1973, p. 6).

A nice illustration of the failure of IIA occurred in the Women's Figure Skating event at the 2002 Winter Olympics in Salt Lake City, Utah. *Sports Illustrated* reporter Brian Cazeneuve described the situation in a Web posting dated February 22, 2003, as follows:

> If you didn't think skating scoring was screwy before Thursday night, you may be asking: How is it that Michelle Kwan was ahead of Sarah Hughes before Irina Slutskaya skated, only to finish behind Hughes after Slutskaya performed? This is the magic of fractured placement, the scoring system that sounds like knee surgery gone awry. . . .

Notice that IIA is, in a sense, saying that the social choice function is derived from one or more voting rules defined on the various subsets v of A. Thus, while transitive rationality is asserting that we are actually in the context of social welfare functions, IIA is asserting that we are actually in the context of voting rules.

We could, at this point, repeat any of the oft-stated rationale for transitive rationality and for independence of irrelevant alternatives. But let's take a different tack and argue on the grounds of naturality. That is, as we observed earlier, many of the common voting systems – plurality voting, the Borda count, the Hare system, etc. – immediately give us examples of both social welfare functions and voting rules. Thus, each of these common voting systems immediately provides two natural examples: one of a social choice function that satisfies P, D, and T, derived from a social welfare function and one that satisfies P, D, and IIA, derived from a voting rule.

In this context, Arrow's theorem asserts that one can't have it both ways – T and IIA are inconsistent in the presence of P and D.

Theorem 1.3.1 (Arrow's Theorem for Social Choice Functions). *If* n *is a positive integer and A is a set of three or more alternatives, then there is no social choice function V for* (A, n) *that satisfies P, D, T, and IIA:*

(P) Pareto: For every (A, n)*-profile* $\boldsymbol{P}$*, every agenda* $v \subseteq A$*, and every pair of alternatives* $x, y \in v$*, if* $x P_i y$ *for every i, then* $y \notin V(\boldsymbol{P})(v)$.

(D) Nondictatorship: There is no i *with the following property: For every* (A, n)*-profile* $\boldsymbol{P}$*, every agenda* $v \subseteq A$*, and every pair of alternatives* $x, y \in v$*, if* $x P_i y$ *for this particular* i*, then* $y \notin V(\boldsymbol{P})(v)$.

(T) Transitive Rationality: There exists a social welfare function V' *for the pair* (A, n) *such that for every* (A, n)*-profile* $\boldsymbol{P}$*, if* $V'(\boldsymbol{P}) = R$*, then, for every agenda* $v \subseteq A$,

$$x \in V(\mathbf{P})(v) \textit{ iff } x \in v \textit{ and for every } y \in v,\ xRy.$$

(IIA) Independence of Irrelevant Alternatives: For every two (A, n)*-profiles* $\boldsymbol{P}$ *and* $\boldsymbol{P}'$*, and every agenda* $v \subseteq A$*, if* $R_i|v = R'_i|v$ *for every i, then*

$$V(\mathbf{P})(v) = V(\mathbf{P}')(v).$$

Arrow actually refers to his result not as a general impossibility theorem, but as a "general *possibility* theorem." The explanation for his usage is implicit in the following quote from Arrow (1950, p. 331): "The aim of the present paper is to show that these difficulties are general. For any method of deriving social choices by aggregating individual preference patterns which satisfies certain natural conditions, *it is possible* [emphasis added] to find individual preference patterns which give rise to a social choice pattern which is not a linear ordering."

Some authors bury the condition of transitive rationality within the formalization by directly working with social welfare functions. Arrow himself tended toward this, stating transitivity rationality early on and then defining D in the context of a social welfare function instead of in the context of a social choice function.

Working directly with social welfare functions allows one to drop T completely, of course, but it also allows one to restate P, D, and IIA in ways that are (perhaps) conceptually simpler. Thus, for example, we have the following version of Arrow's theorem directly in the context of social welfare functions.

Theorem 1.3.2 (Arrow's Theorem for Social Welfare Functions). *If* n *is a positive integer and A is a set of three or more alternatives, then there is no social welfare function V for (A,* n*) that satisfies P, D, T, and PIIA:*

(P) Pareto: For every (A, n*)-profile* $\boldsymbol{P}$*, if* $V(\boldsymbol{P}) = R$*, then for every pair of alternatives* x, y $\in A$*, if* $\mathrm{x P_i y}$ *for every* i*, then* $\mathrm{x}\boldsymbol{P}\mathrm{y}$.

(D) Nondictatorship: There is no i *with the following property: For every (A,* n*)-profile* $\boldsymbol{P}$*, if* $V(\boldsymbol{P}) = R$*, then for every pair of alternatives* x, y $\in A$*, if* $\mathrm{x} P_{\mathrm{i}} \mathrm{y}$ *for this particular* i*, then* $\mathrm{x} P \mathrm{y}$.

(PIIA) Pairwise Independence of Irrelevant Alternatives: For every two (A, n*)-profiles* $\boldsymbol{P}$ *and* $\boldsymbol{P}'$*, if* $V(\boldsymbol{P}) = R$ *and* $V(\boldsymbol{P}') = R'$*, then for every pair of alternatives* x, y $\in A$*,*

$$\text{if } R_{\mathrm{i}}|\{\mathrm{x}, \mathrm{y}\} = R'_{\mathrm{i}}\{\mathrm{x}, \mathrm{y}\} \text{ for every i, then } R|\{\mathrm{x}, \mathrm{y}\} = R'|\{\mathrm{x}, \mathrm{y}\}.$$

Symmetry would seem to demand that there be a version of Arrow's theorem for voting rules, and indeed there is.[6] The version of IIA we use is inspired by the pairwise version of IIA occurring in Arrow's theorem for social welfare functions. Intuitively, that pairwise version says that society's preference for *x* over *y* should be preserved as long as no voter changes the order in which he or

[6] The first version of Arrow's theorem in the context of voting rules seems to be due to Hansson (1969). His version of IIA is, however, awkward at best. The version of Arrow's theorem that we are using here is something we worked out with two undergraduates at Union College, Graham Bryce and William Johnson. It is based on the definition of IIA that we have used in undergraduate classes since the mid-1980s and which appears in Taylor (1995). Similar treatments occur in Denicolo (1985), Beja (1993), and Monjardet (1999).

she ranks x and y relative to each other (by reversing a preference for one over the other, or by introducing an xy-tie or by breaking an xy-tie).

What, then, would it mean, in the context of a voting rule, to say that the outcome of an election reflects a preference that society has for x over y? The only answer to suggest itself here, in the context of a voting rule, is that x should be among the winners and y should not be. But clearly we cannot expect this to be preserved. For example, if all voters were to move some third alternative z to the top of their ballot – thus making z the unique winner in any "reasonable" voting system – then x would not remain a winner, although y would certainly remain a loser. If we were to take a symmetric view of winning and losing, then an analogous argument would say that if all voters were to move some third alternative z to the bottom, then it would be the unique loser. But few would argue that this should be the case in any "reasonable" voting system.

Thus, the most natural version of IIA in the context of voting rules seems to be the one that asserts that if x is a winner and y is a loser and no voter changes the order in which he or she ranks x and y relative to each other (by reversing a preference for one over the other, or by introducing an xy-tie, or by breaking an xy-tie), then y should remain a loser. There are other possibilities as well, but one argument for the naturality of this choice is that the resulting version of Arrow's impossibility theorem for voting rules is equivalent to the version for social welfare functions and the version for social choice functions. For proofs of these equivalencies, see Exercises 7–10.

Theorem 1.3.3 (Arrow's Theorem for Voting Rules). *If* n *is a positive integer and A is a set of three or more alternatives, then there is no voting rule V for* (*A*, n) *that satisfies P, D, and CIIA:*

- *(P) Pareto: For every* (*A*, n)*-profile* ***P*** *and every pair of alternatives* $x, y \in A$*, if* $x P_i y$ *for every* i*, then* $y \notin V(\boldsymbol{P})$.
- *(D) Nondictatorship: There is no* i *with the following property: For every* (*A*, n)*-profile* ***P*** *and every pair of alternatives* $x, y \in A$*, if* $x P_i y$ *for this particular* i*, then* $y \notin V(\boldsymbol{P})$.
- *(CIIA) Choice Independence of Irrelevant Alternatives: For every pair of* (*A*, n)*-profiles* ***P*** *and* ***P***$'$*, and every pair of alternatives* $x, y \in A$*, if* $x \in V(\boldsymbol{P})$ *and* $y \notin V(\boldsymbol{P})$ *and* $R_i|\{x, y\} = R_i'|\{x, y\}$ *for every* i*, then* $y \notin V(\boldsymbol{P}')$.

Proofs of Theorems 1.3.1, 1.3.2, and 1.3.3 (with a little left to the reader) are given in Section 3.4.

In terms of the literature, there are, as one might expect, an abundance of treatments of Arrow's theorem. Two of the most notable book-length treatments are Arrow (1963) and Kelly (1978). Textbook coverage (with proofs) is also widely available, including chapters in Kelly (1987), Saari (1995), and Taylor (1995).

The 1970s seem to have been the heyday for social choice research in the second half of the twentieth century. For references, Kelly (1991) is remarkable. Books on social choice (from a number of different perspectives) include Sen (1970), Fishburn (1973), Feldman (1980b), Riker (1982), Schofield (1985), Campbell (1992), Shepsle and Bonchek (1997), Austen-Smith and Banks (2000), and Arrow, Sen, and Suzumura, (2002). Books on different aspects of voting include Straffin (1980), Nurmi (1987), Saari (1994), Felsenthal and Machover (1998), Taylor and Zwicker (1999), and Saari (2001). An important recent survey is Brams and Fishburn (2002).

At this point, we have the basic definitions of the contexts in which social choice takes place, a little of the history of the field, several informal examples of aggregation procedures, and the statement of Arrow's impossibility theorem in each of our three contexts. In the final section of this chapter, we give precise formulations of twenty voting rules that will all play a role in what follows.

1.4 Twenty Voting Rules

In this section, we work only in the context of linear ballots, and in what follows we give a reasonably broad collection of voting rules that either naturally suggest themselves as real-world voting systems or explicitly arise for one reason or another in the context of manipulation. Each of these procedures is defined for an arbitrary number of voters and alternatives, and thus each voting rule presented also corresponds to one or more social choice functions, a couple of which are of particular relevance to later results on manipulability.

In organizing the presentation of these voting rules, we have grouped them roughly according to similarities in underlying philosophy and the extent to which they satisfy a few desirable properties that have been introduced over the years. The following definition collects six such properties.

Definition 1.4.1. A voting rule V for (A, n) satisfies:

(1) *anonymity* (or is *anonymous*) if it treats all voters the same. More precisely, V is anonymous if, for every permutation σ of the set N of voters,

$$\mathrm{V}(\mathbf{P}) = \mathrm{V}(\sigma(\mathbf{P}))$$

where $\sigma(\mathbf{P}) = <\mathrm{P}_{\sigma(1)}, \ldots, \mathrm{P}_{\sigma(n)}>$.

(2) *neutrality* (or is *neutral*) if it treats all the alternatives the same. More precisely, V is neutral if, for every permutation σ of the set A of alternatives,

$$V(\sigma(P_1), \ldots, \sigma(P_n)) = \sigma(V(P_1, \ldots, P_n))$$

where, for a linear ordering $L = <a_1, \ldots, a_k>$ of A, the sequence $\sigma(L)$ is defined to be $<\sigma(a_1), \ldots, \sigma(a_k)>$.

(3) *monotonicity* (or is *monotone*) if a winner remains a winner when a voter alters his or her ballot by interchanging that winning alternative with the alternative that he or she had ranked immediately above it. For a more precise statement, see Exercise 13.

(4) *Pareto* (or the *Pareto condition*) if no alternative is a winner when there is some other alternative that every voter prefers to it. More precisely, V satisfies Pareto if

$$\forall \mathbf{P}\ \forall x \in V(\mathbf{P})\ \forall y \in A - \{x\}\ \exists i \in N\ x P_i y.$$

(5) *unanimity* if an alternative is the unique winner whenever every voter has it at the top of his or her ballot. More precisely, V satisfies unanimity if

$$\forall \mathbf{P}\ \forall x \in A \text{ if } \forall i \in N \text{ top}_i(\mathbf{P}) = x, \text{ then } V(\mathbf{P}) = \{x\}.$$

(6) *non-imposition* if every alternative in A occurs as the unique winner in at least one election. More precisely, V is non-imposed if

$$\forall x \in A\ \exists \mathbf{P}\ V(\mathbf{P}) = \{x\}.$$

The Pareto condition implies unanimity, which in turn implies non-imposition (Exercise 14).

With these definitions at hand, we now present our first group of voting rules. The six procedures with which we begin all depend on first-place votes only, with (roughly speaking) fewer and fewer first-place votes required to be a winner (more accurately, to be in a set of winners that is a proper subset of A) as we progress from the first to the sixth. All of these procedures are anonymous, neutral, monotone, and satisfy Pareto.

For each of the descriptions, we assume that for some $n \geq 1$ the collection of voters is the set $N = \{1, 2, \ldots, n\}$, the collection of alternatives is the set A, and $\mathbf{P} = (P_1, \ldots, P_n)$ is an arbitrary but linear (A, n)-profile.

(1) The Unanimity Rule: If every voter has the same alternative at the top of his or her ballot, then this alternative is the unique winner; otherwise, all the alternatives tie for the win. More precisely, V is the unanimity rule (often called "rule by unanimity") if, for every $x \in A$,

$$x \in V(\mathbf{P}) \text{ iff } \forall i \in N \ \mathrm{top}_i(\mathbf{P}) = x \text{ or } \forall y \in A \ \exists j \in N \ \mathrm{top}_j(\mathbf{P}) \neq y.$$

(2) The Near-Unanimity Rule: If all but at most one voter have the same alternative at the top of their ballots, then this alternative is the unique winner; otherwise, all the alternatives tie for the win. More precisely, V is the near-unanimity rule if, for every $x \in A$,

$$x \in V(\mathbf{P}) \text{ iff } |\{i \in N: \ \mathrm{top}_i(\mathbf{P}) \neq x\}| \leq 1$$
$$\text{or } \forall y \in A \, |\{i \in N, \ \mathrm{top}_i(\mathbf{P}) \neq y\}| \geq 2.$$

(3) The Plurality Rule: An alternative is a winner if no other alternative has strictly more first-place votes than it has. More precisely, V is the plurality rule (often called "plurality voting") if, for every $x \in A$,

$$x \in V(\mathbf{P}) \ \text{ iff } \ \forall y \in A \, |\{i \in N: \mathrm{top}_i(\mathbf{P}) = x\}| \geq | \ \{i \in N: \ \mathrm{top}_i(\mathbf{P}) = y\}|.$$

(4) The Plurality Runoff Rule: If two or more alternatives are tied for the win with the plurality procedure, then there is a runoff – using the plurality procedure – among those alternatives. If there is a unique plurality winner, then there is a runoff – again using the plurality procedure – between the plurality winner and the alternative (or alternatives) with the second-most first-place votes. More precisely, let V_p denote the plurality rule. Then the plurality runoff rule V is defined as follows:

If $|V_p(\mathbf{P})| = 1$, then $V(\mathbf{P}) = V_p(\mathbf{P} \mid [\{V_p(\mathbf{P})\} \cup V_p(\mathbf{P} \mid \{i : \mathrm{top}_i(\mathbf{P}) \notin V(\mathbf{P})\})])$.
If $|V_p(\mathbf{P})| \geq 2$, then $V(\mathbf{P}) = V_p(\mathbf{P} \mid V_p(\mathbf{P}))$.

(5) The Nomination-With-Second Rule: An alternative is a winner if it has at least two first-place votes (and all alternatives tie for the win if there is no such alternative – which only happens if $n \leq |A|$). More precisely, V is the nomination-with-second rule if

$$x \in V(\mathbf{P}) \ \text{ iff } \ |\{i \in N: \ \mathrm{top}_i(\mathbf{P}) = x\}| \geq 2$$
$$\text{or } \forall y \in A|\{i \in N: \ \mathrm{top}_i(\mathbf{P}) = y\}| \leq 1.$$

(6) The Omninomination Rule: An alternative is a winner if it has at least one first-place vote. More precisely, V is the omninomination rule if

$$V(\mathbf{P}) = \{\mathrm{top}_1(\mathbf{P}), \ldots, \mathrm{top}_n(\mathbf{P})\}.$$

Four of the next five procedures also depend only on first-place votes, but these procedures are, in general, far from anonymous.

(7) Oligarchies: For each set O of voters, we have the voting rule wherein an alternative is a unique winner if every voter in O has that alternative at the top of his or her ballot; otherwise, all the alternatives tie for the win. More precisely, a voting rule V is an oligarchy if there exists a set O of voters such that, for every $x \in \mathrm{A}$,

$$x \in \mathrm{V}(\mathbf{P}) \text{ iff } \forall i \in \mathrm{O},\ \mathrm{top}_i(\mathbf{P}) = x \text{ or } \forall y \in \mathrm{A}\, \exists j \in \mathrm{O}\ \mathrm{top}_j(\mathbf{P}) \neq y.$$

(8) Dictatorships: For each voter, we have the voting rule wherein the election winner is the unique alternative that this voter has at the top of his or her ballot. More precisely, a voting rule V is a dictatorship if there exists $i \in \mathrm{N}$ such that

$$\mathrm{V}(\mathbf{P}) = \{\mathrm{top}_i(\mathbf{P})\}.$$

(9) Duumvirates: For each pair of distinct voters, we have the voting rule wherein the election winners are the alternatives that these two voters have at the top of their ballots. More precisely, a voting rule V is a duumvirate if there exists $\{i, j\} \in [\mathrm{N}]^2$ such that

$$\mathrm{V}(\mathbf{P}) = \{\mathrm{top}_i(\mathbf{P}), \mathrm{top}_j(\mathbf{P})\}.$$

(10) Triumvirates: For each triple of distinct voters, we have the voting rule wherein the election winners are the alternatives that these three voters have at the top of their ballots. More precisely, a voting rule V is a triumvirate if there exists $\{i, j, k\} \in [\mathrm{N}]^3$ such that

$$\mathrm{V}(\mathbf{P}) = \{\mathrm{top}_i(\mathbf{P}), \mathrm{top}_j(\mathbf{P}), \mathrm{top}_k(\mathbf{P})\}.$$

(11) Antidictatorships: For each voter, we have the voting rule wherein the election winner is the unique alternative that this voter has at the bottom of his or her ballot. More precisely, a voting rule V is an antidictatorship if there exists $i \in \mathrm{N}$ such that, if $\mathbf{P}^*$ is the result of turning the ballots in $\mathbf{P}$ upside down (i.e., $x\mathrm{P}_i^* y$ iff $y\mathrm{P}_i x$), then

$$\mathrm{V}(\mathbf{P}) = \{\mathrm{top}_i(\mathbf{P}^*)\}.$$

There is, of course, some overlap in our descriptions. For example, an oligarchy with the set O being a singleton is a dictatorship, and one with the set O being the whole set N is the unanimity rule.

There does not seem to be a common name for what we are calling "the omninomination rule" and the "nomination-with-second rule." We chose the former because this procedure corresponds to a situation in which each voter is allowed to nominate one candidate (pointed out to us by William Zwicker) and the latter based on the observation that it simply captures the seconding aspect of most nomination systems (pointed out to us by Gwendolyn Taylor). The term "duumvirate" is from Feldman (1980a); a dictionary definition of duumvirate is the "union of two men in the same office, or the office, dignity, or government of two men thus associated."

Our third group of voting rules consists of five that involve one-on-one contests between pairs of alternatives based on the ballots cast. As with the voting rules in our first group of five, all of these procedures are anonymous, neutral, and monotone, and – although not all satisfy Pareto – they all satisfy unanimity.

Again, for each of the descriptions, we assume that for some $n \geq 1$ the collection of voters is the set $\mathrm{N} = \{1, 2, \ldots, n\}$, the collection of alternatives is the set A, and $\mathbf{P} = (\mathrm{P}_1, \ldots, \mathrm{P}_n)$ is an arbitrary but linear (A, n)-profile. Additionally, for any two alternatives x and y, we let $\mathrm{W}(x, y, \mathbf{P})$ denote the number of voters who rank x over y on their ballot in $\mathbf{P}$; thus, $\mathrm{W}(x, y, \mathbf{P}) = |\{i \in \mathrm{N}\colon x\mathrm{P}_i y\}|$. Notice that x defeats y in a one-on-one contest based on the ballots cast (in $\mathbf{P}$) iff

$$\mathrm{W}(x, y, \mathbf{P}) > \mathrm{W}(y, x, \mathbf{P}),$$

and x and y tie iff

$$\mathrm{W}(x, y, \mathbf{P}) = \mathrm{W}(y, x, \mathbf{P}).$$

(12) The Condorcet Rule: If there is an alternative that would strictly defeat every other alternative in a one-on-one contest based on the ballots cast, then that alternative is the unique winner; otherwise, all the alternatives tie for the win. More precisely, V is the Condorcet voting rule (often called "the Condorcet voting procedure") if, for every $x \in \mathrm{A}$,

$$x \in \mathrm{V}(\mathbf{P}) \text{ iff } \forall y \in \mathrm{A} - \{x\}\, \mathrm{W}(x, y, \mathbf{P}) > \mathrm{W}(y, x, \mathrm{P})$$

or

$$\forall y \in \mathrm{A}\, \exists z \in \mathrm{A} - \{y\}\, \mathrm{W}(z, y, \mathbf{P}) \geq \mathrm{W}(y, z, \mathbf{P}).$$

Most descriptions of the Condorcet rule simply say there is no winner in the case in which no alternative can strictly defeat every other in a one-on-one contest. We chose to have all the alternatives tie in this case because that seems to be most in keeping with the original spirit (all winning, in some sense, being equivalent to all losing). There are, however, certainly other possibilities that are superior to this in a number of ways.

(13) The Weak Condorcet Rule: If there is at least one alternative that would defeat or tie every other alternative in a one-on-one contest based on the ballots cast, then all such alternatives are winners; otherwise, all the alternatives tie for the win. More precisely, V is the weak Condorcet voting rule if, for every $x \in \mathrm{A}$,

$$x \in \mathrm{V}(\mathbf{P}) \text{ iff } \forall y \in \mathrm{A}\ \mathrm{W}(x, y, \mathbf{P}) \geq \mathrm{W}(y, x, \mathbf{P})$$

or

$$\forall y \in \mathrm{A}\ \exists z \in \mathrm{A}\ \mathrm{W}(z, y, \mathbf{P}) > \mathrm{W}(y, z, \mathbf{P}).$$

(14) Copeland's Rule: An alternative is a winner if no alternative has a strictly better win–loss record, where the win–loss record is calculated as the number of strict wins minus the number of strict losses. More precisely, V is the Copeland rule (often called "the Copeland procedure") if, for every $x \in \mathrm{A}$,

$$x \in \mathrm{V}(\mathbf{P}) \text{ iff } \forall y \in \mathrm{A}, (1) > (2)$$

where

$$(1) = |\{z \in \mathrm{A}: \mathrm{W}(x, z, \mathrm{P}) > \mathrm{W}(z, x, \mathrm{P})\}| - |\{z \in \mathrm{A}: \mathrm{W}(z, x, \mathrm{P}) > \mathrm{W}(x, z, \mathrm{P})\}|$$

and

$$(2) = |\{z \in \mathrm{A}: \mathrm{W}(y, z, \mathrm{P}) > \mathrm{W}(z, y, \mathrm{P})\}| - |\{z \in \mathrm{A}: \mathrm{W}(z, y, \mathrm{P}) > \mathrm{W}(y, z, \mathrm{P})\}|.$$

(15) The Borda Count: An alternative is a winner if no alternative has strictly more total points in one-on-one contests, where a point is gained for an alternative x each time we have $x\mathrm{P}_i z$ for some i and some z. More precisely, the voting rule V is the Borda count if, for every $x \in \mathrm{A}$,

$$x \in \mathrm{V}(\mathbf{P}) \text{ iff } \forall y \in \mathrm{A} \sum\{\mathrm{W}(x, z, \mathbf{P}) : z \in \mathrm{A}\} \geq \sum\{\mathrm{W}(y, z, \mathbf{P}) : z \in \mathrm{A}\}.$$

(16) The Pareto Rule: An alternative is a winner unless there is some alternative that every voter prefers to it. More precisely, the voting rule V is the Pareto procedure if, for every $x \in \mathrm{A}$,

$$x \in \mathrm{V}(\mathbf{P}) \text{ iff } \forall y \in \mathrm{A}\ \exists i \in \mathrm{N}\ x\mathbf{P}_i y.$$

The next procedure also depends on one-on-one contests, but we separate it from the previous five because it is not neutral, and, although it is non-imposed, it (surprisingly – see Exercise 16) does not satisfy Pareto.

(17) The Sequential Pairwise Rule: For each ordering $a_1, \ldots, a_k$ of the alternatives (called, in this context, an *agenda*), we have the voting rule wherein the alternative a_1 is pitted against the alternative a_2, the winner (or both, if they tie) then being pitted against a_3, and so on, with any alternative suffering a loss being eliminated. More precisely, let $\mathrm{V_{WC}}$ denote the weak Condorcet voting rule. Then the voting rule V is the sequential pairwise rule (often called "sequential pairwise voting" or "sequential pairwise voting with a fixed agenda" or "the agenda procedure") if, for every $x \in \mathrm{A}$,

$$\mathrm{V}(\mathbf{P}) = \mathrm{W}_k$$

where the sequence $<\mathrm{W}_1, \ldots, \mathrm{W}_k>$ is defined inductively by setting $\mathrm{W}_1 = \{a_1\}$ and, for $2 \leq r \leq k$,

$$\mathrm{W}_r = \mathrm{V_{WC}}(\mathbf{P}|(\mathrm{W}_{r-1} \cup \{a_r\})).$$

Our final three voting rules are essentially "runoff systems." All are based on the idea of iteratively deleting alternatives that are, in some sense, least preferred. The first such procedure that we present was introduced by Thomas Hare in 1861 (see Section 1.2), and the second such procedure that we present is a variant of the first; it is due to the psychologist Clyde Coombs. Remarkably, neither is monotone (see Exercises 18 and 19).

The Hare procedure and the Coombs procedure are special cases of the general idea of repeatedly using a single procedure to break ties among winners. More precisely, suppose that A is a set of alternatives, n is a positive integer, and V is a voting rule defined not for just (A, n), but for (A$'$, n) for every $\mathrm{A}' \subseteq \mathrm{A}$. Now, for every (A, n) profile $\mathbf{P}$, we can consider the sequence

$$<\mathrm{W}_1, \mathrm{W}_2, \ldots, \mathrm{W}_{|\mathrm{A}|}>$$

where $\mathrm{W}_1 = \mathrm{V}(\mathbf{P})$, $\mathrm{W}_2 = \mathrm{V}(\mathbf{P}|\mathrm{W}_1)$, $\mathrm{W}_3 = \mathrm{V}(\mathbf{P}|\mathrm{W}_2)$, etc.

Notice that

(i) $\mathrm{A} \supseteq \mathrm{W}_1 \supseteq \mathrm{W}_2 \supseteq \ldots \supseteq \mathrm{W}_{|\mathrm{A}|}$, and
(ii) if $\mathrm{W}_j = \mathrm{W}_{j+1}$, then $\mathrm{W}_{j+1} = \ldots = \mathrm{W}_{|\mathrm{A}|}$.

Thus, as we repeatedly apply V to the same set of voters, but with fewer and fewer alternatives remaining, the set of winners eventually stabilizes (perhaps with a single winner, but perhaps with a group remaining tied for the win). We can now let V^* be the voting rule for (A, n) wherein $V^*(\mathbf{P}) = W_{|A|}$.

In the descriptions to follow, we continue to assume that for some $n \geq 1$ the collection of voters is the set $N = \{1, 2, \ldots, n\}$, the collection of alternatives is the set A, and $\mathbf{P} = (P_1, \ldots, P_n)$ is an arbitrary but linear (A, n)-profile.

(18) The Hare Rule: One repeatedly deletes the alternative or the alternatives with the fewest first-place votes, with the last group of alternatives to be deleted tied for the win. More precisely, V is the Hare voting rule (also called "the Hare system" or "the Hare procedure") if $V = V^*_H$ where $V_H(\mathbf{P})$ is the set of all alternatives except those with the fewest first-place votes in **P** (and all tie if all have the same number of first-place votes).

(19) The Coombs Rule: One repeatedly deletes the alternative or the alternatives with the most last-place votes, with the last group of alternatives to be deleted tied for the win. More precisely, V is the Coombs voting rule (also called "the Coombs procedure") if $V = V^*_C$ where $V_C(\mathbf{P})$ is the set of all alternatives except those with the most last-place votes in **P** (and all tie if all have the same number of first-place votes).

(20) The Iterated Plurality Rule: One repeatedly applies the plurality rule to those tied for the win with the plurality rule. More precisely, V is the iterated plurality voting rule if $V = V^*_p$ where $V_p(\mathbf{P})$ is the set of all plurality winners from **P**.

Notice that a voting rule V satisfies Pareto if, for every profile **P**, $V(\mathbf{P}) \subseteq V_{par}(\mathbf{P})$ where V_{par} is the Pareto rule. Similarly, a voting rule satisfies unanimity if it agrees with the unanimity rule for every profile in which the latter procedure has a unique winner.

These twenty voting rules encompass most, but not all, of the important aggregation procedures that arise in the context in which ballots are lists. The most glaring omission might well be a social welfare function known as Kemeny's rule (see Young and Levenglick (1978), Merlin and Saari (2000), and Monjardet (2005)).

We conclude our brief introduction to social choice theory with a result that is of particular relevance to manipulability. The question it addresses is whether or not there exist *any* voting rules that are both natural and democratic and, at the same time, resolute. To formalize this, let's take "natural and democratic" to mean "anonymous, neutral, and satisfies Pareto."

In one sense, the answer to this question is no, there simply aren't any voting rules that are anonymous, neutral, and resolute. That is, the voting rules that we've considered are defined for every set A of alternatives and every n, and when one says the "procedure is resolute" it's reasonable to assume they mean "for every set A of alternatives and for every n, the corresponding voting rule for (A, n) is resolute."

But if $|A| = 3$ and $n = 3$, the ballots in the Condorcet voting paradox are such that any anonymous and neutral voting rule must make all three alternatives tied for the win. The same observation applies to the case where $|A| = 2$ and $n = 2$ and the two voters submit opposite rankings of the two candidates.

This answer, however, is too dismissive. The point is that, in seeking resoluteness, we should fix the size of the set A of alternatives, and then ask for which n (if any) resolute voting rules for (A, n) exist.

In particular, theorems like those of Arrow and of Gibbard and Satterthwaite apply on an "(A, n)-by-(A, n) basis." That is, they are not assertions of the form; "If V is a voting rule that satisfies certain conditions for every A and every n, then certain conclusions hold for every A and every n." Rather, they say: "If V is a voting rule that satisfies certain conditions for some A and some n, then certain conclusions hold for that particular A and that particular n."

As a starting point in asking about resoluteness, consider the trivial case wherein $|A| = 2$. If n is even, then there are no resolute voting rules that are anonymous and neutral, as can be seen by considering a profile in which half the voters rank the alternatives one way and half the voters rank them the opposite way. If n is odd, then majority rule is resolute, anonymous, and neutral. The following theorem provides one generalization of this.

Theorem 1.4.2. *In the context of linear ballots, if* n *is a multiple of some number between 2 and* $|A|$ *inclusive, then every voting rule for* (A, n) *that satisfies anonymity, neutrality, and Pareto fails to be resolute. On the other hand, if* n *is not a multiple of any number between 2 and* $|A|$ *inclusive, then the Hare procedure is a resolute social choice function for* (A, n) *that satisfies anonymity, neutrality, and Pareto.*

Proof: Assume first that $2 \leq k \leq |A|$ and that $n = mk$. Let $A = \{a_1, \ldots, a_k, \ldots, a_{|A|}\}$. Consider the profile **P** in which the n voters are broken into m groups of size k and each group of size k has the following sequence of ballots:

$$
\begin{array}{ccccccc}
a_1 & a_2 & a_3 & \cdot & \cdot & \cdot & a_k \\
a_2 & a_3 & a_4 & \cdot & \cdot & \cdot & a_1 \\
a_3 & a_4 & a_5 & \cdot & \cdot & \cdot & a_2 \\
\cdot & \cdot & \cdot & \cdot & \cdot & \cdot & \cdot \\
\cdot & \cdot & \cdot & \cdot & \cdot & \cdot & \cdot \\
\cdot & \cdot & \cdot & \cdot & \cdot & \cdot & \cdot \\
a_k & a_1 & a_2 & & & & a_{k-1} \\
a_{k+1} & a_{k+1} & a_{k+1} & \cdot & \cdot & \cdot & a_{k+1} \\
\cdot & \cdot & \cdot & \cdot & \cdot & \cdot & \cdot \\
\cdot & \cdot & \cdot & \cdot & \cdot & \cdot & \cdot \\
\cdot & \cdot & \cdot & \cdot & \cdot & \cdot & \cdot \\
a_{|A|} & a_{|A|} & a_{|A|} & \cdot & \cdot & \cdot & a_{|A|}
\end{array}
$$

With any voting rule satisfying Pareto, the winners must be among $a_1, \ldots, a_k$. But then anonymity and neutrality imply that these must tie for the win. Because $k \geq 2$, this shows that resoluteness fails.

Conversely, we show by induction on $|A|$ that, for every n, if n is not a multiple of any number between 2 and $|A|$ inclusive, then the Hare procedure for (A, n) is resolute. We leave to the reader the verification that the Hare system satisfies anonymity, neutrality, and Pareto (regardless of A and n).

If $|A| = 1$, then for every n the Hare system (and everything else) is clearly resolute. So suppose that $|A| > 1$, that n is not a multiple of any number between 2 and $|A|$ inclusive, and that **P** is a linear (A, n)-profile. It can't be the case that all the alternatives have exactly the same number of first-place votes, because that would imply that n is a multiple of $|A|$. Hence, with the Hare system, we delete a proper subset of A in the first stage, and we have a linear (A', n)-profile $\mathbf{P}'$ in which n is not a multiple of any number between 2 and $|A'|$ inclusive. The inductive hypothesis now shows that applying the Hare system to $\mathbf{P}'$ will result in a unique winner. □

There are other voting rules that, like the Hare system, are resolute for every pair (A, n) that is not ruled out by Theorem 1.4.2; see Exercise 28.

1.5 Exercises

(1) [S] Assume that R is a weak ordering of A.

a. Prove that the derived relation P is asymmetric, irreflexive, and transitive.

b. Prove that the derived relation I is reflexive, symmetric, and transitive.

(2) [C] Prove that for every n, if $A = \{a_1, \ldots, a_n\}$, then there exists a linear (A, n)-profile **P** such that $\forall x \in A\ \exists y \in A\ |\{i \in N: yP_ix\}| = n - 1$. (This is the general version of Condorcet's voting paradox.)

(3) [S] Recall that for n voters and n alternatives, Moulin's voting procedure V_M is defined so that $a \notin V_M(\mathbf{P})$ iff $\exists X \subseteq N\ \exists B \subseteq A$, $|X| + |B| > n$ and $\forall i \in X\ \forall b \in B$, bP_ia. Prove that this is equivalent to the following: $a \notin V_M(\mathbf{P})$ iff $\exists Y \subseteq N\ \exists C \subseteq A - \{a\}$, $|Y| + |C| \leq n - 2$ and, if we eliminate the voters in Y and the alternatives in C, then a is on the bottom of the remaining ballots, i.e., $\forall i \in N - Y\ \forall b \in A - C$, bP_ia or $b = a$.

(4) [S] Prove that Moulin's procedure satisfies Pareto, and that for $n = 3$ there is always a winner. Hint: For the first part, Exercise 3 is useful.

(5) [C] Given a linear (A, n)-profile **P**, consider the family $\{B_a: a \in A\}$ of functions where, for $(i, b) \in N \times (A - \{a\})$, we have $B_a(i, b) = 1$ iff aP_ib. The Borda score of an alternative a is the sum, over all voters i in N, of the sum, over all alternatives b in A, of $B_a(i, b)$.
   (a) Write this description of $B(a)$ using sigma notation.
   (b) The Borda score of an alternative a can also be found by calculating a's total score in one-on-one contests with all the other alternatives. Write down this expression in sigma notation (it should differ from what is in part a).
   (c) What third-grade rule of arithmetic implies that the expression from part a is numerically equal to the one from part b?
   (d) Which of the two expressions most directly relates to the calculation of a Borda score by counting the occurrences of other alternatives below a?

(6) [C] Use the following profile **P** to show that the narrow Borda rule can differ from the broad Borda rule. That is, find a subset v of alternatives such that, if V is the Borda rule, then $V(\mathbf{P}|v)$ is not the alternative in v with the highest Borda score from **P**.

P

d	b	c	d
a	c	d	a
b	d	a	c
c	a	b	b

The following two exercises outline a derivation showing that the version of Arrow's theorem for voting rules implies the version for social welfare functions (and thus, because of transitive rationality, the version for social choice functions). We assume that ballots are linear.

(7) [T] Notationally, if **P** is a profile, let $<a_1, \ldots, a_k>$ denote voter *i*'s ballot P_i and, if V is a social welfare function, let $<B_1, \ldots, B_r>$ denote the weak ordering V(**P**) where everything in "block" *j* is tied and above everything in block $j + 1$.

a. Show that if V is a social welfare function for (A, *n*) that satisfies IIA and Pareto, and if we define a voting rule V′ for (A, *n*) by setting V′(**P**) = B_1 (the top block in V(**P**)), then V′ satisfies IIA and Pareto.

b. Suppose V is a social welfare function for (A, *n*) satisfying IIA and suppose that voter *i* is a "top dictator" in the sense that the top block B_1 in V(**P**) is the single element that voter *i* has at the top of his ballot in **P**. Show that voter *i* is, in fact a dictator (i.e., V(**P**) = P_i). Hint: We want to show that $B_j = \{a_j\}$ for every $j = 1, \ldots, k$. For this, assume that **P** shows otherwise and choose the least index *j* such that $B_j \neq \{a_j\}$. Choose $a_{j+p} \in B_j$. What can we conclude if voter *i* moves a_j to the top of his or her ballot?

(8) [S/T] Use the results in Exercise 7 to show that the version of Arrow's theorem for voting rules implies the version for social welfare functions (and thus also implies the version for social choice functions).

The following two exercises outline a derivation showing that the version of Arrow's theorem for social welfare functions implies the version for voting rules. We assume that ballots are linear, and we continue with the notation from Exercise 7.

(9) [T] Every voting rule V for (A, *n*) that satisfies Pareto gives rise to a social welfare function V′ for (A, *n*) in a natural way. Informally, we can describe this derivation of V′ from V as follows. Given an (A, *n*)-profile **P**, we begin by letting the top block B_1 for V′(**P**) be the set V(**P**). To get the second block B_2 for V′(**P**), two ideas suggest themselves: (i) delete the alternatives in V(**P**) and take B_2 to be the resulting set of winners; or (ii) move all the alternatives in V(**P**) to the bottom of everyone's ballot, and again take B_2 to be the resulting set of winners. Both ideas can be repeated to yield B_3, B_4, etc. It turns out that the second idea is the most useful (e.g., it preserves IIA whereas the first doesn't), and our assumption of Pareto ensures that each block is disjoint from the ones that came before it.

For a more precise description of V′, we need a little notation. If P_i is a linear A-ballot, and $B \subseteq A$, then BP_i denotes the linear A-ballot where xBP_iy iff we have $x \notin B$ and $y \in B$ or we have xP_iy with either $x, y \in B$ or $x, y \notin B$. If **P** is a linear (A, *n*)-profile and $B \subseteq A$, then $\mathbf{BP} = <BP_1, \ldots, BP_n>$. Now, given a voting rule V for (A, *n*) that satisfies Pareto and a linear

(A, n)-profile **P**, we can inductively define a sequence of subsets of A by setting $B_1 = V(\mathbf{P})$, and, if $B_1, \ldots, B_j$ have been defined and $\mathbf{B}_1 \cup \ldots \cup B_j \neq A$, then $B_{j+1} = V(\mathbf{B}_j \ldots \mathbf{B}_1\mathbf{P})$. We can now let $V'(\mathbf{P})$ be the weak ordering R given by $x\text{R}y$ iff $x \in B_i$ and $y \in B_j$ and $i \leq j$.

(a) If $A = \{a, b, c, d, e\}$, $B = \{b, d\}$, and $C = \{c\}$, find CBP_1, where P_1 is the ballot that ranks the alternatives in alphabetical order.

(b) If V is the plurality rule, find $V'(\mathbf{P})$ where **P** is the following profile:

P

a	a	a	c	d
b	d	a	d	c
c	b	d	a	b
d	c	c	b	a

(c) Find a profile with three alternatives and three voters that shows that if V is the plurality rule, then V′ is not ranking the alternatives according to the number of first-place votes, and if V is the Borda count, then V′ is not ranking the alternatives according to their Borda scores.

(d) Show that V′ satisfies Pareto. To get started notationally, assume that **P** is a linear (A, n)-profile in which every voter prefers a to b. Let $<B_1, \ldots, B_j>$ be a listing of the "blocks" in the weak ordering $V'(\mathbf{P})$ and assume, for contradiction, that $a \in B_r$ and $b \in B_s$ and $s \leq r$. Let $Q = \mathbf{B}_{s-1} \ldots \mathbf{B}_1\mathbf{P}$.

(e) Show that if V satisfies Pareto and IIA, then V′ satisfies IIA. Again, to get started, assume for contradiction that IIA for V′ fails. Then we can choose two linear (A, n)-profiles **P** and **P′** and two alternatives a and b such that

 (i) $\{i \in N: aP_ib\} = \{i \in N: aP'_ib\}$,

 (ii) a is over b in $V(\mathbf{P})$, and

 (iii) b is over a in $V(\mathbf{P}')$ or they are tied in $V(\mathbf{P}')$.

Let $<B_1, \ldots, B_j>$ again be a listing of the blocks in the weak ordering $V'(\mathbf{P})$ and let $<B'_1, \ldots, B'_l>$ be a listing of the blocks in $V(\mathbf{P}')$. Choose r and s such that $a \in B_r$ and $b \in B_s$.

(10) [S/T] Use the results in Exercise 9 to show that the version of Arrow's theorem for social welfare functions implies the version for voting rules.

(11) [C] Show that "Pareto" cannot be replaced by "non-imposition" in the statement of Arrow's theorem for either voting rules or social welfare functions.

(12) [T] Complete the following to give a direct proof that, with three or more alternatives (and linear ballots), every voting rule V satisfying IIA and Pareto is resolute.

Assume, for contradiction, that V is not resolute. Choose a profile **P** and distinct alternatives a, b, $c \in$ A such that a, $b \in$ V(**P**). Let X $= \{i \in$ N: $a\mathrm{P}_i b\}$ and Y $= \{i \in$ N: $b\mathrm{P}_i a\}$. Consider the profiles $\mathbf{P_1}$, $\mathbf{P_2}$, $\mathbf{P_3}$, and $\mathbf{P_4}$, where we get:

$\mathbf{P_1}$ by moving c to the bottom of every ballot in **P**.

$\mathbf{P_2}$ by moving c up between a and b on every ballot in $\mathbf{P_1}$ that is held by voters in X.

$\mathbf{P_3}$ by moving c over a on the ballots in $\mathbf{P_2}$ held by voters in X.

$\mathbf{P_4}$ by moving c between a and b on the ballots in $\mathbf{P_3}$ held by voters in Y.

Claim 1. $c \notin \mathrm{V}(\mathbf{P_1})$ and $\{a, b\} \subseteq \mathrm{V}(\mathbf{P_1})$.

Hint. Use Pareto and then show that, if $\{a, b\} \not\subseteq \mathrm{V}(\mathbf{P_1})$, going from $\mathbf{P_1}$ to **P** would be a failure of IIA.

Claim 2. $c \notin \mathrm{V}(\mathbf{P_2})$ and $\{a, b\} \subseteq \mathrm{V}(\mathbf{P_2})$.

Claim 3. $c \notin \mathrm{V}(\mathbf{P_3})$ and $\{a, b\} \subseteq \mathrm{V}(\mathbf{P_3})$.

Claim 4. $c \notin \mathrm{V}(\mathbf{P_4})$ and $\{a, b\} \subseteq \mathrm{V}(\mathbf{P_4})$.

(13) [S] Give a precise statement of the definition of monotonicity in Definition 1.4.1. Notationally, if **P** is a profile, $i \in$ N and x, $y \in$ A with $y\mathrm{P}_i x$, let $\mathbf{P}(i, x, y)$ be the profile **Q** (if it exists) such that $\mathbf{P}|\mathrm{N} - \{i\} = \mathbf{Q}|\mathrm{N} - \{i\}$, $\mathrm{P}_i|\mathrm{A} - \{x, y\} = \mathrm{Q}_i|\mathrm{A} - \{x, y\}$, and $y\mathrm{Q}_i x$. Note that such a **Q** exists iff there is no alternative z between x and y on voter i's ballot.

(14) [S] Prove that the Pareto condition implies unanimity and that unanimity implies non-imposition.

(15) [C] For each of the twenty voting rules in Section 1.4, calculate the set of winners using the following profile **P**. For sequential pairwise voting, take the agenda to be alphabetical; for those procedures that are not anonymous, assume the more powerful voters are to the right of the less powerful voters; take the oligarchy to be of size two.

P

a	a	a	c	c	b	e
b	d	d	b	d	c	c
c	b	b	d	b	d	d
d	e	e	e	a	a	b
e	c	c	a	e	e	a

(16) [C] Use the following profile to show that the Condorcet rule, the weak Condorcet rule, and the sequential pairwise rule do not satisfy Pareto:

P

a	c	b
b	a	d
d	b	c
c	d	a

(17) [S] Prove that the Copeland procedure, the Borda count, and the Coombs procedure all satisfy Pareto.

(18) [C] Use the following profile to show that the Hare system is not monotone. Note that there are 13 voters.

5 voters	4 voters	3 voters	1 voter
a	c	b	b
b	b	c	a
c	a	a	c

(19) [C] Show that the Coombs procedure is not monotone. Hint: An appropriate profile can be found somewhere in Chapter 2 (in the context of the Coombs procedure).

(20) [S/C] Say that a voting rule satisfies the "top condition" if an alternative can never be among the winners unless it has at least one first-place vote.
   (a) Prove that if a voting rule satisfies the top condition, then it satisfies Pareto.
   (b) Identify which of the twenty voting rules from Section 1.4 satisfy the top condition for every A and n.

(21) [C] Consider the 20-by-3 chart in which the rows are labeled by the twenty procedures in Section 1.4 and the columns are labeled by "anonymity," "neutrality," and "Pareto." Fill in a "yes" or "no" according to whether the given procedure has the given property.

(22) [S] Suppose that $\text{A} = \{a, b\}$, $n = 2$, and V is a voting rule for (A, n) that is anonymous and neutral. Prove that V is not resolute. In terms of structure, this can be done by a chain:

$$a \in \text{V}(\mathbf{P}) \text{ iff} \ldots \text{iff } b \in \text{V}(\mathbf{P}),$$

where $\mathbf{P}$ is appropriately chosen, as is $\sigma\colon \{a, b\} \to \{a, b\}$ and $\sigma'\colon \text{N} \to \text{N}$.

(23) [S] Suppose that $A = \{a, b, c\}$, $n = 3$, and V is a voting rule for (A, n) that is anonymous and neutral. Prove that V is not resolute.

(24) [C] Suppose $A = \{a, b, c, d\}$ and $n = 3$. Consider the following (A, n)-profile **P**:

P

a	c	b
b	a	d
d	b	c
c	d	a

Show that if the voting rule being used is sequential pairwise voting and you have "agenda-setting power" (i.e., you get to specify the order in which the one-on-one contests will take place); then you can arrange for whichever of the four alternatives you want to be the single winner.

(25) [S/T] Use Arrow's theorem to show that, for every $n \geq 2$, the Condorcet procedure for (A, n) is not resolute if $|A| \geq 3$.

(26) [C/T] Give a direct proof of the result in Exercise 25. (You might want to separately consider the case where n is odd and the case where n is even.)

(27) [C] Show that if $|A| \geq 3$, then the following procedures reduce to only two distinct ones: unanimity, near unanimity, plurality, nomination-with-second, omninomination, duumvirates, Condorcet, and weak Condorcet.

(28) [T] Prove that both the Coombs procedure and the iterated plurality procedure are resolute for every (A, n) not ruled out by Theorem 1.4.2.

(29) [C/S] Show that for $|A| = 3$ and $n = 5$, the Hare, Coombs, and iterated plurality procedures are not the only resolute voting rules for (A, n) that are anonymous, neutral, and satisfy Pareto. Hint: Consider the procedure that asks, of the two top first-place voter getters (if there's a two-way tie), which does better one-on-one against the third. Use the following profile to prove that this isn't just an alternate description of one of the other three procedures in this special case:

P

a	b	c	a	c
b	c	a	b	b
c	a	b	c	a

(30) [T] Complete the following to prove May's theorem (for linear ballots):

Theorem (*May, 1952*). *If $|A| = 2$ and n is odd, majority rule is the only voting rule for (A, n) that is resolute, anonymous, neutral, and monotone.*

Suppose $A = \{a, b\}$, $N = \{1, \ldots, n\}$ where $n \geq 1$, and V is a voting rule for (A, n). Call a set X of voters a "winning coalition for *a*" if $V(\mathbf{P}) = \{a\}$ whenever the voters in X are precisely the ones to rank *a* over *b* on their ballots in **P** (i.e., $aP_i b$ iff $i \in X$). Similarly, we can define what it means to say that X is a winning coalition for *b*. Let

$$W_a = \{X \subseteq N: \text{ X is a winning coalition for } a\}$$

and

$$W_b = \{X \subseteq N: \text{ X is a winning coalition for } b\}.$$

Now, assume additionally that *n* is odd and that V is both resolute and anonymous. Because V is anonymous, whether *a* wins a particular election depends only on the number of voters who rank *a* over *b* on their ballots (as opposed to which particular voters rank *a* over *b*). Thus, we can let W^*_a (respectively: W^*_b) be the set of all numbers *k* between 0 and *n* with the property that $V(\mathbf{P}) = a$ (respectively: $V(\mathbf{P}) = b$) for every profile **P** in which exactly *k* voters rank *a* over *b*. That is,

$$W^*_a = \{k \in N: \ V(\mathbf{P}) = a \text{ whenever } |\{i \in N: \ aP_i b\}| = k\}$$

and

$$W^*_b = \{k \in N: \ V(\mathbf{P}) = b \text{ whenever } |\{i \in N: \ bP_i a\}| = k\}.$$

To complete the proof, establish each of the following claims:

Claim 1: $k \in W^*_a$ iff $n - k \notin W^*_b$.
Claim 2: V is neutral iff $W^*_a = W^*_b$. Hint for the proof: Let $\mathbf{P}^*$ be the result of turning all ballots in **P** upside down. Then V is neutral iff $\forall\sigma$ and $\forall\mathbf{P}$, $V(\sigma(\mathbf{P})) = \sigma(V(\mathbf{P}))$ iff $\forall\mathbf{P}$, $V(\mathbf{P}) = a$ iff $V(\mathbf{P}^*) = b$ iff $\forall k. \ldots$
Claim 3: V is monotone iff $k \in W^*_a$ implies $k + 1 \in W^*_a$, and $k \in W^*_b$ implies $k + 1 \in W^*_b$.
Claim 4: V is majority rule iff $W^*_a = \{k \in N: k > n - k\} = W^*_b$.
Claim 5: If, in addition to being anonymous and resolute, V is neutral and monotone, then $W^*_a = \{k \in N: k > n - k\} = W^*_b$.

2

An Introduction to Manipulability

2.1 Set Preferences and Manipulability

It has long been known that a voter can sometimes achieve a preferred election result by casting a ballot that misrepresents his or her actual preferences. Over a century ago, C. L. Dodgson referred to a tendency of voters to "adopt a principle of voting which makes it more of a game of skill than a real test of the wishes of the electors" (Black, 1958, p. 232). Dodgson went on to say that in his opinion, it would be "better for elections to be decided according to the wishes of the majority than of those who happen to have most skill at the game" (Black, 1958, p. 233).

The most famous manipulability quote in the history of social choice, however, predates Dodgson by a century or so. It was Jean Charles de Borda's famous reply to a colleague who had pointed out to him how easy it was to manipulate his (Borda's) method of marks (i.e., the Borda count). "My scheme," Borda replied, "is only intended for honest men!" (Black, 1958, p. 182).

Alas, the practice of manipulation today is not restricted to men or women we would consider dishonest. For example, in his 1986 book, *The Art of Political Manipulation*, the late William H. Riker (considered by many to be the intellectual founding father of positive political theory) provides a dozen stories that illustrate the extent to which

> politicians are continually poking and pushing the world to get the results they want. The reason they do this is they believe (and rightly so) that they can change outcomes by their efforts. It is often the case that voting need not have turned out the way it did.

This poking and pushing is the issue we now address.

In a study of manipulation of voting systems, there are two rather distinct types of questions. With the first, one begins with an explicitly given aggregation procedure and attempts to find the ways in which a voter can secure a more favorable election outcome by a unilateral change in his or her ballot. With the second, one starts with an explicit notion of what it means for a voter to prefer one outcome to another and attempts to find all the aggregation procedures (of a certain kind) that are manipulable in this sense. The first kind of question is generally felt to be considerably easier than the second,[7] and it is largely what we pursue in this chapter. The remainder of the book is devoted to the second kind of question.

Throughout this chapter, we assume that ballots are linear. Our starting point is to address the question of how one formalizes the notion of manipulability.

Intuitively, a voting system is manipulable if there exists an election in which some voter can secure an outcome that he or she prefers by unilaterally changing his or her ballot. The ballots of the other voters are held fixed. This corresponds to the assumption that this particular voter has complete knowledge of how everyone else voted (or perhaps better: will vote) and can capitalize on this knowledge to secure a better outcome – better, that is, from his or her point of view – by submitting an insincere ballot. We are, by the way, considering only the kind of manipulation that involves a ballot change by a single voter. Group manipulation, also called *coalitional manipulability*, is discussed in Sections 6.3 and 6.4.

More precisely, a voting rule V is manipulable if there are two profiles $\mathbf{P}$ and $\mathbf{P}'$ and a voter i such that $\mathbf{P}|\mathrm{N} - \{i\} = \mathbf{P}'|\mathrm{N} - \{i\}$ and voter i, whose true preferences we take to be P_i, "prefers" $\mathrm{V}(\mathbf{P}')$ to $\mathrm{V}(\mathbf{P})$. If $\mathrm{V}(\mathbf{P}')$ and $\mathrm{V}(\mathbf{P})$ are singletons, say $\mathrm{V}(\mathbf{P}') = \{x\}$ and $\mathrm{V}(\mathbf{P}) = \{y\}$, then there is no doubt what we mean when we say that voter i prefers $\mathrm{V}(\mathbf{P}')$ to $\mathrm{V}(\mathbf{P})$; it simply means that $x\mathrm{P}_i y$. The issue, as we've said earlier, is deciding what it means to say that a voter prefers a set X of alternatives to another set Y.

Almost all the major theorems on manipulability that we present in Chapters 3–8 involve formalizations of set preferences based on one or more of the following ideas:[8]

(1) The case where X and Y are singletons, as we just discussed.
(2) The idea of one set X "weakly dominating" another set Y in the sense that everything in X is at least as good as everything in Y, and something in X is better than something in Y.

[7] I've heard the distinction referred to as low social choice theory versus high social choice theory, but the truth might be that investigators have simply asked (and answered) harder instances of the second question than the first.

[8] There is a vast literature regarding the derivation of subset preferences from preferences over single elements of a set; see, for example, deFinetti (1937), Savage (1954), Fishburn (1986), and Burani and Zwicker (2000). Only some of the resulting ideas have played a prominent role in the study of manipulation of voting systems. Those that have played such a role arise, in one way or another, from an intuition based on the supposition that a single winner will eventually be selected – by a lottery, by a machine, by a person, etc. – from the group initially chosen.

(3) The idea of comparing $\max_i(X, P)$ and/or $\min_i(X, P)$ with $\max_i(Y, P)$ and/or $\min_i(Y, P)$.
(4) The idea of one set X having a higher "expected utility" than another set Y where the calculation is done using some real-valued utility function that represents a voter's preferences and one or more probability functions that provide a measure of the likelihood that a given alternative will ultimately be selected from a given set.

We comment on each of these in turn, and then cull four explicit forms of manipulability from these notions. In the next section of this chapter, we return to our twenty voting rules given in the last chapter and illustrate these different forms of manipulability in that concrete setting.

The first idea above – the case where X and Y are singletons – requires no further comment. The second idea above, that of weak dominance, arises in game theory where one speaks of a strategy for a player as weakly dominating another strategy for that player if the former always yields an outcome at least as good for that player as the latter and sometimes yields an outcome that is strictly better. In point of fact, an election can be thought of as a game in which a strategy for a player (voter) is a choice of ballot, and the outcome of the game is the set of winners in the election.

Thus, we can say that a set X of alternatives is preferred by voter i to a set Y of alternatives in the sense of weak dominance if

$$\forall x \in X\, \forall y \in Y(x R_i y)^9 \quad \text{and} \quad \exists x \in X\, \exists y \in Y(x P_i y).$$

Because ballots are linear, this means that for a set X to weakly dominate a set Y they must have at most one element in common: That is, $\min_i(X, \mathbf{P})\, R_i\, \max_i(Y, \mathbf{P})$ and $X \neq Y$. This, however, allows us to split weak dominance into a "max-version" and a "min-version." In the former, we have

$$\forall x \in X\, \forall y \in Y(x R_i y) \quad \text{and} \quad \max_i(X, \mathbf{P})\, P_i \max_i(Y, \mathbf{P}),$$

and in the latter we have

$$\forall x \in X\, \forall y \in Y(x R_i y) \quad \text{and} \quad \min_i(X, \mathbf{P})\, P_i \min_i(Y, \mathbf{P}).$$

For the third idea, suppose we are given a ballot P_i (in a profile $\mathbf{P}$) that we take to represent the true preferences of voter i. A naïve approach yields four ways to use the $\max_i$ and $\min_i$ functions to assert that one set X of alternatives is preferred by voter i to another set Y of alternatives:

(i) $\max_i(X, \mathbf{P})\, P_i \min_i(Y, \mathbf{P})$
(ii) $\min_i(X, \mathbf{P})\, P_i \max_i(Y, \mathbf{P})$

9 Recall that we are assuming that ballots are linear, so $x R_i y$ means $x P_i y$ or $x = y$.

(iii) $\max_i(X, \mathbf{P})\ P_i\ \max_i(Y, \mathbf{P})$
(iv) $\min_i(X, \mathbf{P})\ P_i\ \min_i(Y, \mathbf{P})$

It turns out that (i) and (ii) are not very satisfactory in terms of giving useful notions of manipulability. In particular, (i) is too weak – $\max_i(X, \mathbf{P})\ P_i\ \min_i(Y, \mathbf{P})$ being neither transitive nor irreflexive on sets with more than one element, and (ii) is strong enough so as to be somewhat redundant – manipulations resulting in $\min_i(X, \mathbf{P})\ P_i\ \max_i(Y, \mathbf{P})$ most commonly achievable only when the sets X and Y can, in fact, be taken to be singletons.

However, there are some reasonably good intuitions behind the use of (iii) and (iv) in manipulability investigations. For example, let's assume that, when the dust settles, society will need to have a single winner, and that this single winner will be selected in some way (randomly, by some committee, etc.) from those tied for the win according to our voting rule.

Now, if a voter is sufficiently optimistic, and if he or she ranks a over b over c over d, then he or she will prefer an election outcome of $\{a, d\}$ to an election outcome of $\{b, c\}$. This is because he or she will assume – optimistically – that a (his or her top choice overall) will result from an election outcome of $\{a, d\}$, while b (his or her second choice overall) will result from an election outcome of $\{b, c\}$. In general, a sufficiently optimistic voter will compare two election outcomes (that is, two sets of alternatives) by asking which has a larger max according to the voter's true preference ranking of the alternatives – that is, by using (iii).

On the other hand, if a voter is sufficiently pessimistic, and if he or she ranks a over b over c over d, then he or she will prefer an election outcome of $\{b, c\}$ to an election outcome of $\{a, d\}$. This is because he or she will assume – pessimistically – that d (his or her worst choice overall) will result from an election outcome of $\{a, d\}$, while c (his or her third choice overall) will result from an election outcome of $\{b, c\}$. In general, a sufficiently pessimistic voter will compare two election outcomes (that is, two sets of alternatives) by asking which has a larger min according to his or her true preference ranking of the alternatives (that is, by using (iv)).

Another way to view this notion of manipulation by an optimist or a pessimist is to return to our example in Chapter 1 wherein we had ten faculty members in an academic department trying collectively to choose from among five candidates for a position in the department. Any of the voting rules that come to mind (plurality, Hare, Borda, etc.) will produce ties upon occasion, and one option is to let the dean break any ties that arise. Here, optimism and pessimism need not be any kind of general state of mind. An optimist is simply a department member who feels that the dean shares his or her values (e.g., the relative importance attached to effective teaching versus a prominent research profile); a pessimist is one who feels just the opposite.

Finally, our fourth notion is based on the idea that one might want to say that a voter prefers a set X of alternatives to a set Y of alternatives if his or her "expected utility" from X is greater than his or her "expected utility" from Y. Although we put expected utility in quotes for good reason, the intuition here is quite clear; the expected utility of a set X of alternatives to a voter should be the sum, over all $x \in$ X, of the product of the following two numbers:

(1) The "value" or "utility" of alternative x to that voter.
(2) The probability with which that voter sees alternative x emerging as the eventual winner from the set X.

For this kind of arithmetic calculation to make sense, we want the "utility" referred to in (1) to be a number. This is achieved if each voter has a so-called *utility function u* mapping the set A of alternatives to the set $\Re$ of real numbers (denoted u: A $\rightarrow \Re$) *that represents his or her preferences* (for individual alternatives) in the sense that for every x, $y \in$ A, $x\text{P}_i y$ iff $u(x) > u(y)$. Additionally, (2) requires that each voter has, for every set X of alternatives, a *probability function on X*, that is, a function p: X $\rightarrow$ [0, 1] such that $\sum\{p(x) : x \in \text{X}\} = 1$. Here again there are two natural ways in which p might arise:

(1) The probability function p might depend on the particular voter and his or her knowledge or suppositions about how ties will ultimately be resolved. In this case, the nature of p might vary from voter to voter and from set to set.
(2) The probability function p might be determined by the procedure itself. For example, if the procedure were to specify that ties must be broken randomly, then we would have $p(x) = 1/|\text{X}|$ for every $x \in$ X.

There are six expected-utility notions of manipulability arising from the situation described in (1), and all of these have combinatorial equivalents that make use of the $\min_i$ and $\max_i$ functions (sometimes in ways that are quite different from what we had above for optimists and pessimists). This material is presented in Section 4.4.

For the moment, however, we want to focus on (2) and the notion of manipulability arising from saying that a set X of alternatives is preferred (or perhaps better, can be preferred) to a set Y of alternatives by voter i if there exists a utility function u representing P_i such that, if $p(x) = 1/|\text{X}|$ for every $x \in$ X, and

$p(y) = 1/|Y|$ for every $y \in Y$, then

$$\sum\{p(x) \cdot u(x) : x \in X\} > \sum\{p(y) \cdot u(y) : y \in Y\}^{10}.$$

This notion was introduced by Feldman, and we illustrate it in the course of proving Theorem 2.3.1 in this chapter. Notationally, if u is a utility function representing P_i, then we let

$$E_{u,i}(X) = \sum\{u(x) : x \in X\}/|X|.$$

Thus, in the special case where $p(x) = 1/|X|$ for every $x \in X$, the "expected utility of X" might also be called the "mean (or average) utility of X."

This completes our discussion of the four fundamental ideas underlying the sense in which a voter might prefer one set of alternatives to another. These ideas, in turn, give rise to the four primary notions of manipulability that we need to analyze the specific voting rules from the last chapter and to summarize, in Section 2.3, some of the main results presented in other chapters. These four notions of manipulability (with comments to follow that allow for slightly finer distinctions) are collected in the following definition.[11]

Definition 2.1.1. In the context of linear ballots, a voting rule is:

(1) *single-winner manipulable* if there exist profiles $\mathbf{P}$ and $\mathbf{P}'$ and a voter i such that $\mathbf{P}|N - \{i\} = \mathbf{P}'|N - \{i\}$ and voter i, whose true preferences we take to be given by his or her ballot in $\mathbf{P}$, prefers the election outcome X from $\mathbf{P}'$ to the election outcome Y from $\mathbf{P}$ in the following sense:

$$X = \{x\} \quad \text{and} \quad Y = \{y\} \quad \text{and} \quad xP_i y.$$

(2) *weak-dominance manipulable* if there exist profiles $\mathbf{P}$ and $\mathbf{P}'$ and a voter i such that $\mathbf{P}|N - \{i\} = \mathbf{P}'|N - \{i\}$ and voter i, whose true preferences we take to be given by his or her ballot in $\mathbf{P}$, prefers the election outcome X from $\mathbf{P}'$ to the election outcome Y from $\mathbf{P}$ in the following sense:

$$\forall x \in X\, \forall y \in Y(xR_i y) \quad \text{and} \quad \exists x \in X\, \exists y \in Y(xP_i y).$$

(3a) *manipulable by optimists* if there exist profiles $\mathbf{P}$ and $\mathbf{P}'$ and a voter i such that $\mathbf{P}|N - \{i\} = \mathbf{P}'|N - \{i\}$ and voter i, whose true preferences we take

[10] If we demanded that the utility function take on only positive real values, then the notion of X being preferred to Y in the sense of expected utility would be unchanged. See Exercise 7.

[11] Attempts to organize the various kinds of manipulability that suggest themselves date back at least to Gärdenfors (1979). Our own experience with this began with several undergraduate theses we supervised, including that of Ryan Kindl and Matthew Gendron. Related material can be found in Bartholdi and Orlin (1991) and Smith (1999).

to be given by his or her ballot in **P**, prefers the election outcome X from **P**′ to the election outcome Y from **P** in the following sense:

$$\max_i(X, \mathbf{P})\ P_i\ \max_i(Y, \mathbf{P})$$

(3b) *manipulable by pessimists* if there exist profiles **P** and **P**′ and a voter i such that $\mathbf{P}|N - \{i\} = \mathbf{P}'|N - \{i\}$ and voter i, whose true preferences we take to be given by his or her ballot in **P**, prefers the election outcome X from **P**′ to the election outcome Y from **P** in the following sense:

$$\min_i(X, \mathbf{P})\ P_i\ \min_i(Y, \mathbf{P})$$

(4) *expected-utility manipulable* if there exist profiles **P** and **P**′ and a voter i such that $\mathbf{P}|N - \{i\} = \mathbf{P}'|N - \{i\}$ and voter i, whose true preferences we take to be given by his or her ballot in **P**, prefers the election outcome X from **P**′ to the election outcome Y from **P** in the following sense: There exists a utility function u representing P_i such that, if $p(x) = 1/|X|$ for every $x \in X$, and $p(y) = 1/|Y|$ for every $y \in Y$, then $\sum\{p(x) \cdot u(x) : x \in X\} > \sum\{p(y) \cdot u(y) : y \in Y\}$; i.e., $E_{u,i}(X) > E_{u,i}(Y)$.

Exercise 2 at the end of the chapter asks for verification that the conditions imposed on X and Y in the definition of weak-dominance manipulability hold iff at least one of the following is true:

(i) $X = \{x\}$ and $Y = \{y\}$ and $x P_i y$
(ii) $\max_i(X, \mathbf{P})\ P_i\ \min_i(X, \mathbf{P})\ R_i\ \max_i(Y, \mathbf{P})$
(iii) $\min_i(X, \mathbf{P})\ R_i\ \max_i(Y, \mathbf{P})\ P_i\ \min_i(Y, \mathbf{P})$

In case (ii) we will use the phrase "max-weak-dominance manipulable," and in case (iii) the phrase "min-weak-dominance manipulable." Exercise 3 asks for a proof that single-winner manipulability implies weak-dominance manipulability, weak-dominance manipulability implies manipulability by optimists or pessimists, and manipulability by optimists or pessimists implies expected-utility manipulability.

We show in Section 4.4 that the set preference notion used in the definition of manipulation by optimists or pessimists is equivalent to some expected-utility notions that differ from what is given above in a couple of important ways. Conversely, the version of expected-utility manipulation given above has a very nice combinatorial equivalent, but not in terms of the $\max_i$ and $\min_i$ functions. Roughly, it says that a set X is preferred to a set Y if there is some alternative z such that voter i has a larger fraction of X than Y at or above z on his or her ballot (see Exercise 4). But let us now turn to the twenty voting rules introduced earlier and see how they stack up in terms of inducing honesty.

2.2 Specific Examples of Manipulation

The four kinds of manipulability that we have at hand, from strongest to weakest, are single-winner manipulability, weak dominance manipulability, manipulability by optimists and/or pessimists, and expected-utility manipulation. Our first result illustrates these varying levels of manipulability with several of the voting rules presented in the last chapter. Notice that the procedures in (i)–(iv) below are anonymous, neutral, monotone, and non-imposed.

Theorem 2.2.1. [12]

(i) For A = {a, b, c, d}*, the Borda count for* (A, 4) *is single-winner manipulable.*
(ii) For A = {a, b, c}*, the plurality rule for* (A, 4) *is weak-dominance manipulable. However, it is never single-winner manipulable.*
(iii) For A = {a, b, c}*, the Condorcet rule for* (A, 3) *is manipulable by both optimists and pessimists. However, it is never weak-dominance manipulable.*
(iv) For A = {a, b, c}*, the nomination-with-second rule for* (A, 4) *is manipulable by optimists (but never by pessimists if* |A| < n)*, and the near-unanimity rule for* (A, 3) *is manipulable by pessimists (but never by optimists).*
(v) For A = {a, b, c}*, the Pareto rule for* (A, 3) *is expected-utility manipulable. However, it is never manipulable by optimists or pessimists.*
(vi) Dictatorships and duumvirates are never expected-utility manipulable.

Let us make a couple of comments before turning to the proof. First, parts (iv) and (vi) of the theorem involve procedures that few would advocate for real-world adoption, but the results are important for the theory in other chapters. Second, of the four "real-world voting systems" in the theorem, we have that the Borda count is (in one sense, at least) most manipulable, followed by the plurality rule, Condorcet's rule, and the Pareto rule in that order.

Each time we need to show that a voting rule is manipulable in some sense, we produce a positive integer n, a set A of alternatives, and two linear (A, n)-profiles **P** and **P**′ that provide an instance of manipulation by voter 1 (who is at the far left) when we regard his or her true preferences to be given by his or her ballot in **P**. We leave the verification that the election winners are what we say they are to the exercises. For notation, we let $\mathrm{F}(x, \mathbf{P}) = |\{i \in \mathrm{N}: \mathrm{Top}_i(\mathbf{P}) = x\}|$, where we think of "F" as standing for "first."

[12] For a finer analysis of the manipulability of these procedures, see the exercises at the end of the chapter. For example, Exercise 8 gives an example of a voting rule that is single-winner manipulable for (A, n) when |A| = 3 and n = 3, and Exercise 10 asks for a proof that the Borda count is not single-winner manipulable when |A| = 3.

Proof: For (i), let $A = \{a, b, c, d\}$, let $n = 4$, and consider the following profiles **P** and **P**′:

P				**P**′			
a	b	d	c	b	b	d	c
b	d	c	a	a	d	c	a
c	c	a	b	d	c	a	b
d	a	b	d	c	a	b	d

If V is the Borda count, then $V(\mathbf{P}) = \{c\}$ and $V(\mathbf{P}') = \{b\}$, so voter 1 has improved the election outcome from being his or her third choice to being his or her second choice.

For (ii), let $A = \{a, b, c\}$, let $n = 4$, and consider the following profiles **P** and **P**′:

P				**P**′			
a	c	c	b	b	c	c	b
b	a	a	a	a	a	a	a
c	b	b	c	c	b	b	c

If V is the plurality rule, then $V(\mathbf{P}) = \{c\}$ and $V(\mathbf{P}') = \{b, c\}$, so voter 1 has improved the election outcome from being his or her third choice to being his or her second and third choices. This shows that the plurality rule is max weak-dominance manipulable. The plurality rule can also be min weak-dominance manipulable, depending on the number of alternatives and the number of voters; see Exercise 13.

For the second claim, we will show that, with the plurality rule, no voter can ever simultaneously improve the min and the max of the set of winners. From this it follows that the plurality rule is not single-winner manipulable. In fact, we'll show that if $\mathbf{P}|N - \{i\} = \mathbf{P}'|N - \{i\}$, then either $V(\mathbf{P}) \subseteq V(\mathbf{P}')$ or $V(\mathbf{P}') \subseteq V(\mathbf{P})$.

Assume for contradiction that $\mathbf{P}|N - \{i\} = \mathbf{P}'|N - \{i\}$ and that we can choose $x \in V(\mathbf{P}) - V(\mathbf{P}')$ and $y \in V(\mathbf{P}') - V(\mathbf{P})$. Without loss of generality, assume that yP_ix, and let $F(x, \mathbf{P}) = k$. Notice that because x is not at the top of voter i's ballot, we also have $F(x, \mathbf{P}') \geq k$. Because $x \in V(\mathbf{P})$ and $y \notin V(\mathbf{P})$, we know that $F(y, \mathbf{P}) \leq k - 1$. But now, because $\mathbf{P}|N - \{i\} = \mathbf{P}'|N - \{i\}$, we know that $F(y, \mathbf{P}') \leq k$. It now follows that because $y \in V(\mathbf{P}')$, we have $x \in V(\mathbf{P}')$ because it also has at least k first-place votes in $\mathbf{P}'$, and this is the desired contradiction.

For (iii), let $A = \{a, b, c\}$, let $n = 3$, and consider the following profiles **P** and **P**′:

	P			**P**′	
a	b	c	a	b	c
c	c	a	b	c	a
b	a	b	c	a	b

If V is the Condorcet rule, then $V(\mathbf{P}) = \{c\}$ and $V(\mathbf{P}') = \{a, b, c\}$, so voter 1 has improved the max of the election outcome from being his or her second choice to being his or her first choice. This shows that the Condorcet rule is manipulable by optimists. For the proof that it is also manipulable by pessimists, see Exercise 15.

For the second claim, assume that $\mathbf{P}|N - \{i\} = \mathbf{P}'|N - \{i\}$, that $V(\mathbf{P}) = Y$, that $V(\mathbf{P}') = X$, and that X weakly dominates Y with respect to P_i, which we take to be voter i's true preferences. Notice first that if Y and X are singletons, say $Y = \{y\}$ and $X = \{x\}$, then we must have xP_iy for X to weakly dominate Y. But this is impossible, because then y would still defeat x one-on-one after voter i's ballot change.

It thus follows that one of X and Y is a singleton, and the other is the whole set A. If $X = \{x\}$, then x must be at the top of voter i's ballot in **P** in order to have X dominate Y. But $V(\mathbf{P}) = A$, and so x was not a Condorcet winner. Clearly, no change in voter i's ballot can covert his or her top choice from not being a Condorcet winner to being a Condorcet winner. Similarly, if $Y = \{y\}$, then y must have been at the bottom of voter i's ballot in **P** in order to have X dominate Y. But then y will remain a Condorcet winner no matter how voter i changes his or her ballot. With the Condorcet rule, it also turns out that a voter can never simultaneously improve the max and the min of the set of winners; see Exercise 24.

For (iv), let $A = \{a, b, c\}$, let $n = 4$, and consider the following profiles **P** and **P**′ (which are the same as in (ii)):

	P				**P**′		
a	c	c	b	b	c	c	b
b	a	a	a	a	a	a	a
c	b	b	c	c	b	b	c

If V is the nomination-with-second rule, then $V(\mathbf{P}) = \{c\}$ and $V(\mathbf{P}') = \{b, c\}$, so voter 1 has improved the election outcome from being his or her third choice to being his or her second and third choices. This shows that the nomination-with-second rule is manipulable by optimists (in fact, max weak-dominance manipulable).

For the second claim regarding the nomination-with-second rule, we have $|A| < n$, so there is always at least one alternative with at least two first-place votes. So the min for voter i is either his or her top choice, or it is at the top of two other voter's ballots, in which case he or she can never make it a loser.

For the near-unanimity rule, let $A = \{a, b, c\}$, let $n = 3$, and consider the following profiles **P** and **P**′:

	P			**P**′	
a	b	c	b	b	c
b	a	a	a	a	a
c	c	b	c	c	b

If V is the near-unanimity rule, then $V(\mathbf{P}) = \{a, b, c\}$ and $V(\mathbf{P}') = \{b\}$, so voter 1 has improved the min of the election outcome from being his or her third choice to being his or her second choice. This shows that the near-unanimity rule is manipulable by pessimists.

To show that the near-unanimity rule is not manipulable by optimists, notice that the max for voter i is either his top choice, or it is some single alternative at the top of every other voter's ballots, in which case he or she can never change the election outcome.

For (v), let $A = \{a, b, c\}$, let $n = 3$, and consider the following profiles **P** and **P**′:

	P			**P**′	
a	a	c	a	a	c
b	c	b	c	c	b
c	b	a	b	b	a

If V is the Pareto rule then $V(\mathbf{P}) = \{a, b, c\}$ and $V(\mathbf{P}') = \{a, c\}$. Now let u be any utility function representing voter 1's preferences in **P** with the average of $u(a)$ and $u(c)$ greater than $u(b)$. For definiteness, let's take $u(a) = 18$, $u(b) = 9$ and $u(c) = 6$. Then the expected utility from $\{a, b, c\}$ is

$$(1/3) \cdot 18 + (1/3) \cdot 9 + (1/3) \cdot 6 = 6 + 3 + 2 = 11,$$

and the expected utility from $\{a, c\}$ is

$$(1/2) \cdot 18 + (1/2) \cdot 6 = 9 + 3 = 12.$$

This shows that the Pareto rule is expected-utility manipulable.

It's easy to see that the Pareto rule can't be manipulated by optimists, since $\text{top}_i(\mathbf{P}) \in \text{V}(\mathbf{P})$ for every $i \in \text{N}$. To see that it can't be manipulated by pessimists takes a little more work. Let $y = \min_i(\text{Y})$ where $\text{Y} = \text{V}(\mathbf{P})$. If y is at the bottom of voter i's ballot, then y will certainly remain a winner no matter how voter i changes his or her ballot. So we can assume that there are alternatives $z_1, \ldots, z_k$ that voter i has below y on his or her ballot and that are non-winners. For voter i to make y a loser, he or she must do it by raising at least one of the z's over y so that, when this is done, every voter will have that z over y.

Choose z_1 to be such that voter i has z_1 below y, but every other voter has z_1 above y. Because z_1 is a non-winner (being below y on voter i's ballot), we can choose some z_2 such that every voter has z_2 over z_1, and, in particular, every voter except voter i has z_2 above y. But because y is a winner, voter i must have z_2 below y. But now z_2 has the same properties as did z_1, and we can continue to produce z_3, z_4, etc., forever.

Finally the statement in (vi) is trivial, and this completes the proof of Theorem 2.2.1. □

In addition to the Borda count, eight of our other voting rules are also single-winner manipulable. The following theorem gives seven; finding the eighth is left to the reader (Exercise 27).

Theorem 2.2.2. *For each of the following, there exists an* $\text{n} \geq 1$ *and a set A of alternatives such that the voting rule for (A,* n*) is single-winner manipulable:*

(1) The plurality runoff rule
(2) The weak Condorcet rule
(3) Copeland's rule
(4) The sequential pairwise rule
(5) The Hare system
(6) The Coombs rule
(7) The iterated plurality rule

Proof: In each of the seven cases, we again produce a positive integer n, a set A of alternatives, and two linear (A, n)-profiles $\mathbf{P}$ and $\mathbf{P}'$ that provide an instance of single-winner manipulation by voter 1 (who is at the far left) when we regard his or her true preferences to be given by his or her ballot in **P**.

(1) The plurality runoff rule: Let A = $\{a, b, c\}$, let $n = 5$, and consider the following profiles **P** and **P**′:

P					**P**′				
a	a	c	c	b	b	a	c	c	b
b	b	a	a	c	a	b	a	a	c
c	c	b	b	a	c	c	b	b	a

If V is the plurality runoff rule, then V(**P**) = $\{c\}$ and V(**P**′) = $\{b\}$, so voter 1 has improved the election outcome from being his or her third choice to being his or her second choice.

(2) The weak Condorcet rule: Let A = $\{a, b, c, d\}$, let $n = 4$, and consider the following profiles **P** and **P**′:

P				**P**′			
a	c	b	d	b	c	b	d
b	a	d	c	a	a	d	c
c	b	c	a	d	b	c	a
d	d	a	b	c	d	a	b

If V is the weak Condorcet rule, then V(**P**) = $\{c\}$ and V(**P**′) = $\{b\}$, so voter 1 has improved the election outcome from being his or her third choice to being his or her second choice.

(3) Copeland's rule: Let A = $\{a, b, c, d, e\}$, let $n = 4$, and consider the following profiles **P** and **P**′:

P				**P**′			
a	c	a	d	c	c	a	d
b	e	e	b	a	e	e	b
c	d	d	e	b	d	d	e
d	b	c	c	e	b	c	c
e	a	b	a	d	a	b	a

If V is Copeland's rule, then V(**P**) = $\{d\}$ and V(**P**′) = $\{c\}$, so voter 1 has improved the election outcome from being his or her fourth choice to being his or her third choice.

(4) The sequential pairwise rule: Let A = $\{a, b, c\}$, let $n = 3$, and consider the following profiles **P** and **P**′:

P			**P**′		
a	b	c	b	b	c
b	c	a	a	c	a
c	a	b	c	a	b

If V is the sequential pairwise rule with the ordering of the alternatives being abc, then V(**P**) = $\{c\}$ and V(**P**′) = $\{b\}$, so voter 1 has improved the election outcome from being his or her third choice to being his or her second choice.

(5) The Hare system: Let A = $\{a, b, c, d\}$, let $n = 5$, and consider the following profiles **P** and **P**′:

P					**P**′				
a	b	c	c	d	b	b	c	c	d
b	a	b	b	b	a	a	b	b	b
c	c	a	a	c	c	c	a	a	c
d	d	d	d	a	d	d	d	d	a

If V is the Hare system, then V(**P**) = $\{c\}$ and V(**P**′) = $\{b\}$, so voter 1 has improved the election outcome from being his or her third choice to being his or her second choice.

(6) The Coombs rule: Let A = $\{a, b, c\}$, let $n = 5$, and consider the following profiles **P** and **P**′:

P					**P**′				
a	b	b	a	a	a	b	b	a	a
b	c	c	c	c	c	c	c	c	c
c	a	a	b	b	b	a	a	b	b

If V is the Coombs rule, then V(**P**) = $\{c\}$ and V(**P**′) = $\{a\}$, so voter 1 has improved the election outcome from being his or her third choice to being his or her first choice.

(7) The iterated plurality rule: See Exercise 26.

This completes the proof of Theorem 2.2.2. □

The manipulability of the last four of our voting rules – the unanimity and nomination rules and the oligarchies and triumvirates – is left to the reader (See Exercises 29–32).

2.3 Summary of the Main Results

All four notions of manipulability – single-winner manipulation, weak dominance manipulation, manipulation by optimists or pessimists, and expected-utility manipulation – arise in one or more of the general theorems that we elsewhere present in this book. These results all assert that a large class of voting systems are susceptible to manipulation in that particular sense. By way of summary, we give the statement (but not the proof) of some of those theorems here.

Single-winner manipulation is a very strong notion indeed, and it is precisely the notion addressed by the seminal Gibbard–Satterthwaite theorem. Unfortunately, the class of voting rules that it identifies as being susceptible to this kind of manipulation – that is, the non-imposed, resolute voting rules that are not dictatorships – omits, for example, nineteen of the twenty voting rules that we presented in the last chapter, at least for some choices of A and n.

One can extend the applicability of the Gibbard–Satterthwaite theorem in two different but related ways: One can change the voting rules so that they become resolute, or one can change the theorem so that it applies to non-resolute procedures. Both approaches, it turns out, involve the same idea – using a linear ordering of the alternatives (perhaps one of the ballots, perhaps a fixed "absentee ballot") as a tie-breaker. Unfortunately, using a random device as a tie-breaker leaves the Gibbard–Satterthwaite theorem inapplicable, because the resulting procedure is not a function.

We record the Gibbard–Satterthwaite theorem here in both its resolute and non-resolute forms. The resolute version reappears (with proof) as Theorems 3.1.2 and Corollary 3.1.12. For the equivalence of the non-resolute version, see Corollary 3.1.13.

Theorem 2.3.1. *For every set A of three or more alternatives and every* $n \geq 1$*, every resolute voting rule for* (A, n) *that is non-imposed and not a dictatorship is single-winner manipulable.*

Equivalently, with the same assumptions on A and n*, a (not necessarily resolute) voting rule that is non-imposed and not a dictatorship is manipulable in the sense that there exist profiles* **P** *and* **P′** *and a voter* i *such that* $\mathbf{P}/N - \{i\} = \mathbf{P}'/N - \{i\}$ *and voter* i*, whose true preferences we take to be given by his or* *her ballot in* $\boldsymbol{P}$*, prefers the election outcome X from* $\boldsymbol{P}'$ *to the election outcome Y from* $\boldsymbol{P}$ *in the following sense:*

$$max_i(X - Y, \boldsymbol{P})P_i\, min_i(Y, \boldsymbol{P}) \quad or \quad max_i(X, \boldsymbol{P})P_i\, min_i(Y - X, \boldsymbol{P}).$$

If we had simply said "$\max_i(X, \mathbf{P})\ P_i\ \min_i(Y, \mathbf{P})$," then the corresponding not-necessarily-resolute version above would still have been strong enough to imply the resolute version of the Gibbard–Satterthwaite theorem. However, the relation given by $\max_i(X, \mathbf{P})\ P_i\ \min_i(Y, \mathbf{P})$ is reflexive on non-singleton sets, so *every* non-resolute social choice function is manipulable in this sense by simply choosing an election with more than one winner and then having a voter (any voter) make absolutely no change at all in his or her ballot.

Limitations such as these seem to be the price we pay for dealing with a notion of manipulability as strong as single-winner manipulability. The plurality rule, after all, is not manipulable in this sense. (Exercise 33 asks the reader to reconcile this fact with the opening paragraph of the preface.) Our second notion, weak-dominance manipulability, is also quite strong, but there are several results in the social choice function context showing that it is achievable for a large class of procedures.

We need a couple of definitions pertaining to a social choice function V. First, V is *quasitransitive* if, for every profile $\mathbf{P}$ and for every triple $\{x, y, z\} \in [A]^3$, if $V(\mathbf{P})(\{x, y\}) = \{x\}$ and $V(\mathbf{P})(\{y, z\}) = \{y\}$, then $V(\mathbf{P})(\{x, z\}) = \{x\}$. Second, V is *pairwise non-imposed* if for every $(x, y) \in A \times A$, there exists a profile $\mathbf{P}$ such that $V(\mathbf{P})(\{x, y\}) = \{x\}$. Third, V is a *pairwise oligarchy* if there exists a set O of voters such that for every profile $\mathbf{P}$ and for every $\{x, y\} \in [A]^2$,

$$V(\mathbf{P})(\{x, y\}) = \begin{cases} \{x\} & \text{if } \forall i \in O\ xP_i y \\ \{y\} & \text{if } \forall i \in O\ yP_i x \\ \{x, y\} & \text{otherwise} \end{cases}$$

The following result (Barberá, 1977a and Kelly, 1977) reappears (with proof) as Theorem 5.1.14.

Theorem 2.3.2. *For every set A of three or more alternatives and every* $n \geq 1$*, every social choice function for* (A, n) *that is quasitransitive, pairwise non-imposed, and not a pairwise oligarchy is weak-dominance manipulable on some two-element agenda.*

In order to state the main result regarding manipulation by optimists or pessimists, we need one definition: A voter i is said to be a *nominator* for a voting rule V if $top_i(\mathbf{P}) \in V(\mathbf{P})$ for every profile $\mathbf{P}$. Thus, for example, every voter is a nominator for both the Pareto rule and the omninomination rule. The following result is due to Duggan and Schwartz (1993 and 2000) and reappears (with proof) as Theorem 4.1.2.

Theorem 2.3.3. *For every set A of three or more alternatives and every* n ≥ 1*, every voting rule for* (A, n) *that is non-imposed and has no nominators is manipulable by either optimists or pessimists.*

It follows from the Duggan–Schwartz theorem that, among all voting rules that are anonymous, non-imposed, and not manipulable by optimists or by pessimists, the omninomination rule V is the most discriminating in the sense that, for every profile **P**, we have $\mathrm{V}(\mathbf{P}) \subseteq \mathrm{V}'(\mathbf{P})$ for every member V′ of the class.

Finally, for expected-utility manipulation we have the following theorem of Feldman (1979a). It reappears (with proof) as Theorem 4.3.2.

Theorem 2.3.4. *For every set A of three or more alternatives and every* n ≥ 1*, every voting rule for* (A, n) *that is non-imposed and neither a dictatorship nor a duumvirate is expected-utility manipulable.*

This completes our summary of the main results to follow. We conclude this chapter with an isolated look at a very different kind of manipulability.

2.4 Agenda Manipulability and Transitive Rationality

Manipulation really comes in two flavors, explicitly articulated by Riker (1982, p. 137).

> If we assume that society discourages the concentration of power [thus ruling out dictatorships, for example], then at least two methods of manipulation are always available, no matter what method of voting is used: First, those in control of procedures can manipulate the agenda (by, for example, restricting alternatives or by arranging the order in which they are brought up). Second, those not in control can still manipulate outcomes by false revelation of values.

Ballot manipulation is what we have been doing, and what the rest of this book is about. But in this brief section, we give an elaboration of Kelly's reference (Kelly, 1978, p. 79) to a weakened version of transitive rationality known as path independence as "another kind of strategy-proofness, dealing not with manipulation of preferences, but with manipulation of agendas." First, we need a definition of what it means to say that a social choice function is subject to agenda-manipulation.

With a social choice function, it certainly makes sense to ask if a voter's thwarted preference for x versus y – thwarted in that y actually wins over x as things now stand – might be overcome if he or she could change (or could have changed) the agenda by adding some new alternatives to it or by deleting some other alternatives (other than x and y, that is) from it. Fishburn (1973, pp. 7 and 8) says that "we may consider a maneuver in which an alternative is legally placed in nomination not because its sponsors think it has any chance of being elected but because they feel that its introduction will increase the chance of the election of their favored alternative." These considerations yield the following.

Definition 2.4.1. A resolute social choice function V is *agenda manipulable* if there exists a profile $\mathbf{P}$, two agendas v and v', two alternatives x and y in $v \cap v'$, and a voter i such that $x\mathrm{P}_i y$, $\mathrm{V}(\mathbf{P})(v) = y$, and $\mathrm{V}(\mathbf{P})(v') = x$. If V is not agenda manipulable, then V is *agenda non-manipulable*.

Intuitively, we are thinking of v as having been the original agenda, and voter i's strict preference for x versus y being originally thwarted by y's ability to win over x with the agenda v. However, if voter i were to have agenda-setting power, he or she could – while retaining y in the agenda – switch to v' and obtain x as the winner.

Riker (1986, p. 148) gives a real-world example of exactly this kind of agenda manipulation. He describes how Thomas B. Reed, a Republican member of congress in the late 1800s frustrated his opponents' attempt to expand the agenda from {yes, no} to {yes, no, abstain}. Had his opponents succeeded, the outcome of the election discussed there might well have changed.

Regarding agenda manipulation, we have the following observation.

Theorem 2.4.2. *In the linear-ballot context, a resolute social choice function is agenda non-manipulable iff it satisfies Arrow's condition of transitive rationality.*

Proof: Assume first that V is a resolute social choice function that satisfies transitive rationality, and let V′ be the social welfare function that gives rise to V. Let $\mathbf{P}$ be a profile and assume that $\mathrm{V}'(\mathbf{P}) = \mathrm{R}$. Now, if we have two agendas v and v' with x and y in both, then $\mathrm{V}(\mathbf{P})(v) = y$ implies that y is ranked strictly higher than x according to the relation R. Hence, we can't have $\mathrm{V}(\mathbf{P})(v') = x$, and this shows that V is agenda non-manipulable.

Conversely, assume that V is a resolute social choice function that is agenda non-manipulable. We will produce the social welfare function V′ from which V arises. Let $\mathbf{P}$ be any profile, and set $\mathrm{V}'(\mathbf{P}) = \mathrm{R}$, where R is defined as follows:

$$\text{For } x, y \in \mathrm{A},\ x\mathrm{R}y \text{ iff } \mathrm{V}(\mathbf{P})(\{x, y\}) = x.$$

For the rest of the proof, the profile **P** is fixed, so we will write "V(v)" instead of "V(**P**)(v)."

Claim 1. R is transitive.

Proof. Suppose that xRy and yRz, and assume, for contradiction, that xRz fails. Thus, $V(\{x, y\}) = x$ and $V(\{y, z\}) = y$, but $V(\{x, z\}) \neq x$. Because V is resolute, this means that $V(\{x, z\}) = z$. Again, because V is resolute, we know that $V(\{x, y, z\})$ is either x, y, or z. Without loss of generality, assume that $V(\{x, y, z\}) = x$.[13]

Fix a voter i, and suppose first that $x\text{P}_i z$ (that is, suppose that voter i ranks x over z on his or her ballot). Then, if $v = \{x, z\}$, voter i's preference for x over z is initially thwarted by the fact that $V(\{x, z\}) = z$. But if voter i has agenda-setting power, he or she can add y to the agenda v, obtaining $v' = \{x, y, z\}$, and thus benefiting from the fact that $V(\{x, y, z\}) = x$.

On the other hand, if $z\text{P}_i x$, then we can reverse what we did in the previous paragraph. That is, we take v to be the agenda $\{x, y, z\}$, and note that voter i's preference for z over x is thwarted by the fact that $V(\{x, y, z\}) = x$. But now, if voter i has agenda-setting power, he or she can delete y from the agenda v, obtaining $v' = \{x, z\}$, and thus benefiting from the fact that $V(\{x, z\}) = z$. Because ballots are linear, this completes the proof of Claim 1.

Claim 2. R is complete.

Proof. If xRy fails then $V(\{x, y\}) \neq x$. Because V is resolute, this means $V(\{x, y\}) = y$, and so yRx, as desired.

Claim 3. For any agenda v,

$$V(v) = x \quad \text{iff } x \in v \quad \text{and} \quad \forall y \in v\ [y \neq x \Rightarrow x\text{R}y].$$

Proof. Assume first that $V(v) = x$ and that $y \in v$ with $y \neq x$. If xRy fails, then $V(\{x, y\}) = y$. Fix a voter i, and suppose first that $x\text{P}_i y$. Then, if $v = \{x, y\}$, voter i's preference for x over y is initially thwarted by the fact that $V(\{x, y\}) = y$. But if voter i has agenda-setting power, he or she can expand the agenda to v, and thus benefit from the fact that $V(v) = x$. Similarly if $y\text{P}_i x$, then, voter i's preference for y over x is thwarted by the fact that $V(v) = x$. But if voter i has agenda-setting power, he or she can shrink the agenda to $\{x, y\}$, and thus benefit from the fact that $V(\{x, y\}) = y$.

[13] Let's see why we lose no generality. Suppose that we had $V(\{x, y, z\}) = y$ instead. Then we could replace the first sentence in the proof of the claim with the following equivalent version: "Suppose that yRz and zRx, and assume, for contradiction, that yRx fails." We would then have y playing the same role that x plays in the proof above. A similar remark applies if $V(\{x, y, z\}) = z$.

Conversely, assume that $x \in v$ and $\forall y \in v\ [y \neq x \Rightarrow x\mathrm{R}y]$. Assume that $\mathrm{V}(v) = z$ for some $z \neq x$. But then, by the previous paragraph, we'd have that $\forall y \in v\ [y \neq z \Rightarrow z\mathrm{R}y]$. In particular, we'd have that $x\mathrm{R}z$ and $z\mathrm{R}x$, so $\mathrm{V}(\{x, z\}) = x$ and $\mathrm{V}(\{x, z\}) = z$; a clear contradiction. This completes the proof of Theorem 2.4.2.[14] □

We could easily restate Definition 2.4.1 so that it speaks of a social choice functon V being agenda-manipulable *by a voter* i. One could then say that a social choice fucntion is *weakly agenda-manipulable* if it is agenda-manipulable by at least one voter i, and *strongly agenda-manipulable* if it is agenda-manipulable by every voter i.

If we did this, what we are calling "agenda-manipulable" would then correspond to "weakly agenda-manipulable" in this sense. But it is easy to see that the proof of Theorem 2.4.2 goes through with either notion, and so the two are equivalent – something that is quite trivial to see in its own right.

This completes our discussion of agenda manipulability. Throughout the remainder of the book, we consider only the kind of manipulation in which a single voter achieves a preferred election outcome by submitting a disingenuous ballot. We begin in Chapter 3 with the easiest context: Resolute social choice functions.

2.5 Exercises

(1) [S] Suppose that **P** is a linear (A, n)-profile giving the true preferences of voter i. Let P be the relation defined on sets of alternatives by

$$\mathrm{XPY} \text{ iff } \max_i(\mathrm{X}, \mathbf{P})\ \mathrm{P}_i\ \min_i(\mathrm{Y}, \mathbf{P}).$$

Prove that R is neither irreflexive nor transitive.

(2) [S] Suppose that **P** is a linear (A, n)-profile giving the true preferences of voter i, and that X and Y are sets of alternatives. Prove that voter i prefers X to Y in the sense of weak-dominance manipulation iff one of the following three conditions holds:

(i) $\mathrm{X} = \{x\}$ and $\mathrm{Y} = \{y\}$ and $x\mathrm{P}_i y$.
(ii) $\max_i(\mathrm{X}, \mathbf{P})\ \mathrm{P}_i\ \min_i(\mathrm{X}, \mathbf{P})\ \mathrm{R}_i\ \max_i(\mathrm{Y}, \mathbf{P})$.
(iii) $\min_i(\mathrm{X}, \mathbf{P})\ \mathrm{R}_i\ \max_i(\mathrm{Y}, \mathbf{P})\ \mathrm{P}_i\ \min_i(\mathrm{Y}, \mathbf{P})$.

[14] The relation R defined in the proof of Theorem 2.4.2 is known in the literature as the "base relation." For a great deal of related material, the reader can start with Sen's 1971 paper (available in Sen, 1982) "Choice Functions and Revealed Preferences."

(3) [S] Prove that single-winner manipulability implies weak dominance manipulability, that weak dominance manipulability implies manipulability by optimists or pessimists, and that manipulability by optimists or pessimists implies expected-utility manipulation. Hint: For the first two implications, use Exercise 2.

(4) [S] Suppose that **P** is a profile giving the true preferences of voter i, and that X and Y are sets of alternatives. For every $z \in A$, let $G_i(z) = \{x \in A: x R_i z\}$. Prove that X is preferred to Y in the sense of expected-utility if there exists some $z \in A$ such that

$$|X \cap G_i(z)|/|X| > |Y \cap G_i(z)|/|Y|.$$

(It turns out that the converse is also true, and follows from results later in the book.)

(5) Prove that if $a P_i b P_i c$, and $X = \{a, c\}$ and $Y = \{a, b, c\}$, then X can be preferred to Y in terms of expected utility and Y can be preferred to X in terms of expected utility. (Hint: Use Exercise 4.)

(6) Suppose P_i is a linear ballot in which $a_1 P_i a_2 P_i \ldots P_i a_k$. Let $X = \{a_2\}$ and let $Y = \{a_1, a_2, \ldots, a_k\}$. Prove that there exists a utility function u representing P_i such that $E_{u,i}(X) > E_{u,i}(Y)$. Do the same for $X = \{a_2\}$ and $Y = \{a_1, \ldots, a_k\}$.

(7) [S] Suppose P_i is a linear ballot, X, $Y \subseteq A$ and u: $A \to \Re$ is a utility function representing P_i such that $E_{u,i}(X) > E_{u,i}(Y)$. Let r be an arbitrary real number and define u': $A \to \Re$ by $u'(x) = u(x) + r$. Prove that u' also represents P_i and $E_{u',i}(X) > E_{u',i}(Y)$.

(8) [C/S] Let $A = \{a, b, c\}$ and $n = 3$. Let V be the voting rule for (A, n) wherein the winner is the alternative with the greatest total of first and second-place votes, with ties broken, where possible, by the number of first-place votes. (This is actually an example of what Young (1975) calls a "tie-breaking scoring system.")
   (a) Show that V is not resolute, but that it is single-winner manipulable.
   (b) Show that V satisfies Pareto (and note that it also satisfies anonymity, neutrality, and monotonicity).

(9) [C/S] Let $A = \{a, b, c\}$ and $n = 3$. Let V be the voting rule for (A, n) wherein the winner is the alternative with the greatest number of first- and second-place votes. Prove that V is weak-dominance manipulable, but not single-winner manipulable.

(10) [S] Show that if $|A| = 3$, then for every n, the Borda count for (A, n) is not single-winner manipulable.

(11) [C] Show that if $|A| \geq 4$, then for every n, the Borda count for (A, n) is single-winner manipulable.

(12) [C/S] Show that if $|A| \geq 3$, then the plurality rule for (A, n) is weak-dominance manipulable iff $n \geq 4$.
(13) [C] Show that If $|A| = 3$ and $n \geq 3$, then the plurality rule for (A, n) is min-weak dominance manipulable iff $n \neq 3, 4$ or 6.
(14) [C] Show that if $|A| \geq 3$ and $n \neq 2$ or 4, then the plurality rule for (A, n) is manipulable by pessimists. (Hint: Build on Exercise 10.)
(15) [C] Show that if $|A| \geq 3$ and $n \geq 3$, the Condorcet rule is manipulable by both optimists and pessimists. (Hint: Find suitable profiles for $n = 3$ and $n = 4$, and then show that if such profiles exist for n, then they also exist for $n + 2$.)
(16) [S] Show that the Condorcet rule is never single-winner manipulable.
(17) [C/S] Show that if $|A| \geq 3$ and $n \geq 3$, the nomination-with-second rule is manipulable by pessimists iff $|A| \geq n$, and manipulable by optimists iff $n \geq 4$.
(18) [C] Show that if $|A| \geq 3$ and $n \geq 3$, the near-unanimity rule is manipulable by pessimists.
(19) [S] Prove that if a voter prefers a to b to c, then there exists a utility function realizing these preferences such that his or her expected utility from $\{a, b, c\}$ is higher than from $\{a, c\}$ and another utility function realizing these preferences such that his or her expected utility from $\{a, c\}$ is higher than from $\{a, b, c\}$.
(20) [C] Show that if $|A| \geq 3$ and $n \geq 3$, the Pareto for (A, n) is expected-utility manipulable. (Hint: Use Exercise 16.)
(21) [S] Show that if $|A| = 3$ and $n \geq 3$, the weak Condorcet rule is not single-winner manipulable.
(22) [C/S] Show that if $|A| \geq 4$ and $n \geq 3$, the weak Condorcet rule is single-winner manipulable iff n is even.
(23) [C] In the proof of Theorem 2.2.1, prove that the election winners are what they are advertised to be.
(24) [S] Prove that with the Condorcet rule, one cannot simultaneously improve the min and the max of the set of winners.
(25) [C] In the proof of Theorem 2.2.2, prove that the election winners are what they are advertised to be.
(26 [C] Prove that if $|A| = 3$ and $n = 5$, the iterated plurality rule is single-winner manipulable. (Hint: Suitable profiles occur in the proof of Theorem 2.2.2.)
(27) [C] Find the eighth single-winner manipulable voting rule not covered by Theorems 2.2.1 and 2.2.2.
(28) [S] A voter is a "dummy" for a voting rule if his or her ballot has no effect on the outcome of the election. For example, a dictatorship can be

thought of as arising from the unanimity rule with one voter by the addition of dummies. Prove that manipulability is unaffected by the addition or subtraction of dummies.

(29) [C/S] Show that the unanimity rule for $n \geq 2$ can't be manipulated by an optimist. Do the same for an oligarchy O, assuming that $|O| \geq 2$. Exercise 28 is relevant.

(30) [S] Prove that the omninomination rule and a triumvirate can't be manipulated by optimists or pessimists. Exercise 28 is relevant.

(31) [C] Prove that the omninomination rule and a triumvirate are expected-utility manipulable if $|A| \geq 3$ and $n \geq 3$.

(32) [S] Prove that a duumvirate is not expected-utility manipulable.

(33) [S] Reconcile the fact that the plurality rule is not single-winner manipulable with the first paragraph of the preface.

(34) [T] Use the Gibbard–Satterthwaite theorem to show that, for every $n \geq 2$, the Condorcet rule for (A, n) is not resolute if $|A| \geq 3$. (Compare this with one of the exercises in Chapter 1.)

(35) [S/T] Prove that if $|A| = 2$ and n is arbitrary, then a resolute voting rule for (A, n) is manipulable iff it is monotone.

(36) [T] Prove that for two alternatives and an odd number of voters, majority rule is the only voting rule that is resolute, anonymous, neutral, and non-manipulable. (An exercise in Chapter 1 is relevant.)

3
Resolute Voting Rules

3.1 The Gibbard–Satterthwaite Theorem

For the sake of keeping this chapter relatively self-contained, we begin by restating the definition of manipulability in our present context of resolute voting rules for three or more alternatives. As we are considering only resolute procedures, there is no need to use the phrase "single-winner" when speaking of manipulability.

Definition 3.1.1. In the context of linear or non-linear ballots, a resolute voting rule V is *manipulable* if there exists a profile $\mathbf{P} = (R_1, \ldots, R_n)$, which we think of as giving the true preferences of the n voters, and another ballot Q_i, which we think of as a disingenuous ballot from voter i such that, letting $\mathbf{P}' = (R_1, \ldots, R_{i-1}, Q_i, R_{i+1}, \ldots, R_n)$, we have:

$$V(\mathbf{P}')P_i V(\mathbf{P})$$

If V is not manipulable, then V is *non-manipulable*.

A natural question at this stage is the following: For a given set of alternatives and a given set of voters, exactly which resolute voting rules are non-manipulable? A moment's thought suggests two special cases in which non-manipulable, resolute voting rules certainly exist:

(1) If there is only one voter, then we can take the alternative at the top of his or her ballot as the winner.
(2) If there are only two alternatives and an odd number of voters, then we can use majority rule.

In fact, both ideas apply in the general case where there are several alternatives and several voters. That is, we could implement idea (1) by choosing a particular voter and ignoring all other voters except this one. Or we could implement idea (2) by choosing a particular pair of alternatives and ignoring all other alternatives except these two.

In the context of linear ballots, *all* non-manipulable resolute voting rules arise as natural generalizations of these two observations. However, we put off this discussion until we present some characterization theorems in Chapter 7, and turn instead to the main result of this chapter (indeed, the main result in the study of manipulability of voting systems).

In terms of history, it was in the early 1970s that the philosopher Allan Gibbard and the economist Mark Satterthwaite independently established the theorem bearing their names. Gibbard's result appeared in his article "Manipulation of Voting Schemes: a General Result," *Econometrica 41* (1973) 587–601; Satterthwaite included the result in his 1973 Ph.D. thesis, written at the University of Wisconsin, and he published it later in an article entitled "Strategy-proofness and Arrow's Conditions: Existence and Correspondence Theorems for Voting Procedures and Social Welfare Functions," *Journal of Economic Theory* 10 (1975), 187–217.

The version of the Gibbard–Satterthwaite theorem we state here assumes Pareto, but we later give the easy argument showing that the assumption of non-imposition suffices. Both Gibbard and Satterthwaite handled the case where ties in the ballots are allowed, as we do in Section 3.

Theorem 3.1.2 (The Gibbard–Satterthwaite Theorem for Linear Ballots). *In the context of linear ballots, if* n *is a positive integer and A is a set of three or more alternatives, then any resolute voting rule for* (*A*, n) *that is non-manipulable and that satisfies Pareto is a dictatorship.*

The more important a theorem is, the harder we should look for explanations of why it is true. And the Gibbard–Satterthwaite theorem is extremely important. Thus we find, as we should, a number of different proofs in the literature over the past three decades. Examples include Gärdenfors (1977), Schmeidler and Sonnenschein (1978), Feldman (1979c), Barberá (1983), Benoit (2000), Arunava Sen (2001), and Taylor (2002). For other book-length treatments of manipulability, see Moulin (1983 and 1985) and Riker (1986).

Before beginning the proof of Theorem 3.1.2, let's set the stage by talking about the overall strategy, which is, in fact, the same strategy underlying most proofs of Arrow's impossibility theorem. Assume, then, that V satisfies Pareto and is non-manipulable. Our goal is to "find" the voter *i* who is, in fact, the dictator for V.

The first part of the strategy is to see if we can restrict ourselves to talking about *pairs* of alternatives. For example, the definition of a dictatorship speaks about the whole set A of alternatives and the effect of a dictator arranging all of these in some order on his or her ballot. But what can we say if we only know that for two particular alternatives – call them a and b – the dictator ranks a over b? We certainly cannot conclude that a is the winner. But we *can* conclude that b is a non-winner. Moreover, if a voter had this power for every pair of alternatives, then he or she would, of necessity, be a dictator.

Now, let's consider the other extreme. Instead of looking at a single voter (a dictator, say), let's look at the whole set N of voters. Suppose they all rank a over b on their ballots. What can we conclude? Again, we certainly cannot conclude that a is the winner. But, if Pareto holds, then we can conclude that b is a non-winner.

This suggests the following. If X is a set of voters, call X a "dictating set" if it has the following property: For every profile **P** and every pair a, b of alternatives, if everyone in X ranks a over b, then $V(\mathbf{P}) \neq b$.

Our overall task is now set. We are beginning with the knowledge that the set N itself is a dictating set – this is precisely what Pareto asserts – and we want to find a single voter i such that $\{i\}$ is a dictating set, because this is equivalent to asserting that voter i is a dictator. One way to accomplish this, as we show in the following paragraph, is to prove a lemma that guarantees that if a dictating set is split into two pieces, then one of the two pieces is again a dictating set.

Such a lemma will immediately yield the desired result, because we know that N is a dictating set, and so one exists, and thus there must be one of minimal size, which must then be a singleton by the proposed lemma. Moreover, if the result we are trying to prove is true, then the lemma must also be true because the presence of a dictator ensures that the dictating sets are precisely the sets to which he or she belongs. Hence, our attempts to prove the lemma are not doomed before we start unless the theorem itself is false.

So how do we prove such a lemma? The key is in obtaining one more refinement. Being a dictating set means that for every pair of alternatives, a certain thing happens: Namely, if everyone ranks one of the alternatives over another, the latter is not the winner. It certainly makes sense to ask of a *given pair* a, b of alternatives if this same thing happens. Formalizing this yields the following.

Definition 3.1.3. If X is a set of voters, and a and b are two distinct alternatives in the set A, then *X can use* a *to block* b, denoted aXb, if, for every profile **P** in which all the voters in X rank a over b on their ballots, $V(\mathbf{P}) \neq b$. The set X is a *dictating set* if aXb for every distinct pair a, b of alternatives in A.

Definition 3.1.3 takes place in the context of a fixed resolute voting rule V, and our notation "aXb" and terminology "dictating set" could have been chosen to reflect that dependence on V. For example, we could have chosen to write

"aXb (mod V)" and to speak of a "dictating set for V." The approach in Definition 3.1.3, however, should cause no confusion.

Up to this point, we could as easily have been talking about proving Arrow's impossibility theorem as the Gibbard–Satterthwaite theorem. To bring manipulability into the argument, we could go after the desired lemma directly or – and this is the route we will pursue – we could identify an election-theoretic consequence of manipulability that is combinatorially easier to apply. Consider the following.

Definition 3.1.4. A resolute voting rule V satisfies *down-monotonicity* provided that, for every profile **P**, if **P**′ is the profile obtained from **P** by having one voter move one losing alternative down one spot on his or her ballot, then $\mathrm{V}(\mathbf{P}') = \mathrm{V}(\mathbf{P})$.

Notice that if V satisfies down-monotonicity and $\mathrm{V}(\mathbf{P}) = x$, then $\mathrm{V}(\mathbf{P}') = x$ whenever **P**′ is derived from **P** by having *several* voters move *several* losing alternatives down *several* slots on their ballots. This observation will be important in applying down-monotonicity. One can also show that down-monotonicity implies monotonicity (see Exercise 2).

In our proof of the Gibbard–Satterthwaite theorem, the only direct appeal that we will make to manipulability is in the following lemma.

Lemma 3.1.5. *Every resolute voting rule that is non-manipulable satisfies down-monotonicity.*

Proof: Assume that down-monotonicity fails for the resolute voting rule V. Then there exist two profiles **P** and **P**′ and an alternative y such that:

(1) In **P**, voter i ranks y directly over x, $\mathrm{V}(\mathbf{P}) = w$, and $w \neq y$ (that is, y is the losing alternative that voter i will be moving down).
(2) **P**′ differs from **P** only in that voter i has interchanged the position of x and y on his or her ballot and yet $\mathrm{V}(\mathbf{P}') = v$ for some $v \neq w$.

The situations described in (1) and (2) are pictured below:

P		**P**′	
ballot i	winner	ballot i	winner
y		x	
x	$w \neq y$	y	$v \neq w$

Assuming (1) and (2), we will show that the system can, in fact, be manipulated.

Case 1: v is over w on voter i's ballot in **P**.

In this case, we can regard voter i's ballot in **P** as giving his or her true preferences. Thus, if he or she submits the sincere ballot **P**, w is the winner, although he or she prefers v to w. But if he or she submits a disingenuous ballot (the one in **P**$'$), then v is, in fact, the winner, and he or she prefers v to w according to his or her true preferences given in **P**.

Case 2: w is over v on voter i's ballot in **P**$'$.

In this case, we can regard voter i's ballot in **P**$'$ as giving his or her true preferences. Thus, if he or she submits the sincere ballot **P**$'$, v is the winner, although he or she prefers w to v. But if he or she submits a disingenuous ballot (the one in **P**), then w is, in fact, the winner, and he or she prefers w to v, according to his or her true preferences given in **P**$'$.

Case 3: Otherwise.

In this case, w is over v on voter i's ballot in **P**, and v is over w on voter i's ballot **P**$'$. But this means that $w = y$ and $v = x$ contracting our assumption that $y \neq w$. □

To motivate the next lemma, suppose that we have a set X of voters and two distinct alternatives a and b. Let's think about how we would verify that aXb. If we appeal directly to Definition 3.1.3, then we have to examine *every* profile in which all the voters in X rank a over b, and check to see that b is, in fact, a non-winner in each of these elections.

It would be nice if, in the definition aXb, we could replace the initial universal quantifier ("for every profile **P**. . . . ") with an existential one ("there exists a profile **P**. . . . "). This is, in fact, possible, if we also make two other changes simultaneously: We must require that everyone not in X place b over a, and instead of asserting that V(**P**) $\neq b$, we must get the stronger conclusion that V(**P**) $= a$ (which implies that V(**P**) $\neq b$ because we are assuming resoluteness). These comments are formalized in Lemma 3.1.6 ("the existence lemma") below.

For the remaining five lemmas, we assume that V is a resolute voting rule for (A, n) in the context of linear ballots, and that V satisfies down-monotonicity and Pareto.

Lemma 3.1.6 (The Existence Lemma). *Suppose that X is a set of voters and that* a *and* b *are two distinct alternatives in A. Then, in order to show that* aXb*, it suffices to produce one profile* **P** *for which:*

(1) everyone in X ranks a *over* b*,*
(2) everyone else ranks b *over* a*, and*
(3) $V(\mathbf{P}) = \mathrm{a}$.

Proof: Suppose we have such a profile **P** but aXb fails. Then we also have a profile **P**′ in which everyone in X ranks a over b, and V(**P**′) = b. With **P**′, some voters not in X might also rank a over b, but because we are assuming down-monotonicity, we can obtain a new profile **P**″ by having every such voter move the losing alternative a down below b, and still have V(**P**″) = b. Thus, with both **P** and **P**″, (1) and (2) above hold, with V(**P**) = a and V(**P**″) = b.

Now, choose an alternative c that is distinct from a and b and have every voter move c to the bottom of his or her ballot in both profiles. By down-monotonicity, the winner is still a in the first election and still b in the second. Now choose an alternative d that is distinct from a, b, and c (if there is one) and do the same thing. Continuing this, we eventually get two elections having identical sequences of ballots, with alternative a winning the first and alternative b winning the second. This contradiction completes the proof. □

The next lemma is difficult to motivate directly, but it is precisely the statement whose proof is provided by using the Concorcet voting paradox ballots in our present context.

Lemma 3.1.7 (The Splitting Lemma). *Suppose X is a set of voters and that* a, b, *and* c *are distinct alternatives in A. Assume also that* aXb *and that X is partitioned into disjoint subsets Y and Z (one of which may be empty). Then either* aYc *or* cZb.

Proof: Consider any profile **P** in which every voter in Y has a first, b second, and c third; every voter in Z has c first, a second, and b third; and everyone else (i.e., those voters in N–X) has b first, c second, and a third (with all other alternatives below these). We can picture these ballots as follows:

P

ballots of voters in Y	ballots of voters in Z	ballots of voters in N–X
a	c	b
b	a	c
c	b	a
.	.	.
.	.	.
.	.	.

By Pareto, V(**P**) ∈ $\{a, b, c\}$ (everyone, for example, prefers a to d, so V(**P**) $\neq d$). Moreover, V(**P**) $\neq b$ because aXb by assumption, and everyone in X = Y ∪ Z ranks a over b. But now, Lemma 3.1.6 (the existence lemma) shows that if V(**P**) = a, then aYc, and if V(P) = c, then cZb. □

Lemma 3.1.8. *Suppose X is a set of voters and that* a, b, *and* c *are three distinct alternatives in A. Then*

(1) if aXb, *then* aXc, *and*
(2) if aXb, *then* cXb.

Proof: Notice that in Lemma 3.1.7, we allowed Y or Z to be the empty set. Because Pareto holds, we never have $a\emptyset b$. Conclusions (1) and (2) now follow immediately from Lemma 3.1.7 by letting Y = N and Z = Ø, and then Y = Ø and Z = N. □

Lemma 3.1.9. *Suppose X is a set of voters and that* aXb *holds for some* a *and* b. *Then X is a dictating set.*

Proof: Assume x and y are distinct alternatives. We show that $x\mathrm{X}y$ must hold.

Case 1. $y \neq a$
Because $a\mathrm{X}b$ and $y \neq a$, we know by Lemma 3.1.8 (1) that $a\mathrm{X}y$. Because $x \neq y$, we can now apply Lemma 3.1.8 (2) to get $x\mathrm{X}y$, as desired.
Case 2. $x \neq b$
Because $a\mathrm{X}b$ and $x \neq b$ we know by Lemma 3.1.8 (2) that $x\mathrm{X}b$. Because $y \neq x$, we can now apply Lemma 3.1.8 (1) to get $x\mathrm{X}y$, as desired.
Case 3. $y = a$ and $x = b$
Because A has three or more elements, we can choose c distinct from a and b. Now, because $a\mathrm{X}b$, we know by Lemma 3.1.8 (1) that $a\mathrm{X}c$, and by Lemma 3.1.8 (2) that $b\mathrm{X}c$. A final application of Lemma 3.1.8 (1) shows that $b\mathrm{X}a$, and so $x\mathrm{X}y$, as desired. □

Lemma 3.1.9 allows us to conclude that a set X of voters is a dictating set if $a\mathrm{X}b$ for even *one* pair of alternatives a and b. With this observation and our discussion preceding Definition 3.1.3 (formalized as Lemma 3.1.11 below), we can conclude the proof of Theorem 3.1.2 with the following two lemmas.

Lemma 3.1.10. *If X is a dictating set and X is split into disjoint subsets Y and Z, then either Y is a dictating set or Z is a dictating set.*

Proof: If a, b, and c are any three distinct alternatives, then we know $a\mathrm{X}b$ because X is a dictating set. But now, by Lemma 3.1.7, either $a\mathrm{Y}c$ or $c\mathrm{Z}b$. By Lemma 3.1.9, Y is a dictating set in the former case, and Z is a dictating set in the latter. □

Lemma 3.1.11. *If X is a dictating set, then there exists a voter* i ∈ X *such that* {i} *is a dictating set. In particular, because N is a dictating set (by Pareto), there exists a voter who is a dictator for V.*

Proof: This is immediate from Lemma 3.1.10 and the discussion preceding Definition 3.1.3. □

These lemmas complete the proof of the Gibbard–Satterthwaite theorem in the context of linear ballots. In the next section, we handle the case of non-linear ballots, but before moving on, we state (as corollaries) two equivalent versions of the Gibbard–Satterthwaite theorem. The first replaces Pareto with the weaker assumption of non-imposition (and recasts the result to make it a characterization theorem). The second illustrates what the Gibbard–Satterthwaite theorem yields in the context of non-resolute voting rules.

Corollary 3.1.12. *In the context of linear ballots and three or more alternatives, a voting rule is non-manipulable, non-imposed, and resolute iff it is a dictatorship.*

Proof: A dictatorship is clearly non-manipulable, non-imposed, and resolute. For the converse, it suffices to show that if V is non-manipulable, non-imposed, and resolute, then V satisfies Pareto.

Suppose, for contradiction, that Pareto fails and let **P** be a profile in which every voter ranks a over b, but V(**P**) $= b$. Because V is non-manipulable, it satisfies down-monotonicity, so we lose no generality in assuming that every voter has plunged every alternative other than a and b below b. Hence, every voter has a at the top of his or her ballot.

Because V is non-imposed, we can choose a profile **P**′ with V(**P**′) $= a$. Now, one by one, replace the ballots in **P** with those of **P**′ (so voter 1's ballot in **P** is replaced by his or her ballot in **P**′, then voter 2's ballot in **P** is replaced by his or her ballot in **P**′, and so on). At some point, the election outcome becomes a for the first time. If that last change of ballot is voter i, then we can assume his or her original ballot – with a on top – represents his or her true preferences. But then his or her change to a disingenuous ballot changes the result from something other than his or her top choice to his or her top choice a. This is an instance of manipulation, and completes the proof. □

Corollary 3.1.13. *In the context of linear ballots and three or more alternatives, every non-imposed (not necessarily resolute) voting rule for* (A, n) *that is not a dictatorship is manipulable in the sense there exist profiles* **P** *and* **P**′ *and a* *voter* i *such that* $\boldsymbol{P}|N - \{i\} = \boldsymbol{P}'|N - \{i\}$ *and voter* i, *whose true preferences we take to be given by his ballot in* $\boldsymbol{P}$, *prefers the election outcome X from* $\boldsymbol{P}'$ *to the election outcome Y from* $\boldsymbol{P}$ *in the following sense:*

$$max_i(X - Y, \boldsymbol{P})P_i\, min_i(Y, \boldsymbol{P}) \text{ or } max_i(X, \boldsymbol{P})P_i\, min_i(Y - X, \boldsymbol{P}).$$

Proof: Assume that V is a non-imposed (not necessarily resolute) voting rule that is not a dictatorship. Fix a linear ordering L of the set A of alternatives, and let V′ be the resolute voting rule in which V′(**P**) is the (unique) L-largest element of V(**P**). Clearly, V′ is non-imposed and not a dictatorship, and so V′ is manipulable in the Gibbard–Satterthwaite sense. Thus, there exists an election in which some voter who has x over y on his or her ballot can unilaterally change the outcome from y to x by submitting a disingenuous ballot. But then, with V, the set Y of winners with his or her sincere ballot included y, and the set X of winners with his disingenuous ballot included x. Moreover, we can't have both x and y in X ∩ Y. If $x \notin$ Y, then

$$\max_i(\mathrm{X} - \mathrm{Y}, \mathbf{P})\, \mathrm{R}_i\, x\, \mathrm{P}_i\, y\, \mathrm{R}_i \min_i(\mathrm{Y}, \mathbf{P}).$$

If $y \notin$ X, then

$$\max_i(\mathrm{X}, \mathbf{P})\, \mathrm{R}_i\, x\, \mathrm{P}_i\, y\, \mathrm{R}_i \min_i(\mathrm{Y} - \mathrm{X}, \mathbf{P}).$$

□

3.2 Ties in the Ballots

Both Gibbard's and Satterthwaite's original proofs handled the case where ties in the ballots are allowed. Later, it was realized that one could first prove the result for linear ballots, and then derive the general case from this, as we now do. First, we need a definition in the context of non-linear ballots.

Definition 3.2.1. In the context of non-linear ballots, a voter is a *weak dictator* for a resolute voting rule if the unique winner of every election is one of the alternatives that he or she has tied for top on his or her ballot. V is a *weak dictatorship* if there exists a voter who is a weak dictator for V.

Theorem 3.2.2 (The Gibbard–Satterthwaite Theorem for Non-Linear Ballots). *In the context of non-linear ballots, if* n *is a positive integer and A is a set of three or more alternatives, then any resolute voting rule for* (*A*, n) *that is non-imposed and non-manipulable is a weak dictatorship.*

Proof: We derive Theorem 3.2.2 from Corollary 3.1.12 as follows. Let V′ be the restriction of V to linear ballots. Then clearly V′ is non-manipulable because

V is. We also claim that V′ satisfies unanimity, as we now show. So suppose for contradiction that **P** is a profile in which every voter has a linear ballot with the same alternative x at the top and that V(**P**) $\neq x$. Choose a profile **P**′, with ties, perhaps, such that V(**P**′) $= x$. Now, starting with the profile **P** of linear ballots, change them one by one into **P**′ until x becomes the winner. This last change of ballot (by voter i, say) represents a manipulation of V because voter i, whose true preferences, we assume, are given by his or her linear ballot in **P** with x at the top, changed ballots and secured a win for his or her top choice x.

It now follows from Corollary 3.1.12 that V′ is a dictatorship. Thus, we can assume that if the ballots are linear, then voter i's top choice is the unique winner. We claim that voter i is a weak dictator for V.

Suppose voter i is not a weak dictator for V. Then there exists a profile **P** for which V(**P**) $= x$, but x is not among the alternatives that voter i has tied on top of his or her ballot. One by one, move x up on every other ballot in **P** so that x is alone at the top. Then x remains the winner, or undoing such a move would represent a successful manipulation by that voter. Similarly, we can one by one break all the ties in these other ballots and still have x the winner, lest undoing such a change would yield a voter his or her top choice x, thus again constituting manipulation. Finally, we can break all the ties in voter i's ballot, in which case some alternative that was in his or her top block becomes the winner. Thus, if the original ballot represented voter i's true preferences, then he or she has gained by the disingenuous breaking of ties in the ballot. This completes the proof. □

The converse of Theorem 3.2.2 fails in the following sense: It is not true, in the context of non-linear ballots, that every weak dictatorship is non-manipulable; see Exercise 3. However, some weak dictatorships are non-manipulable in the context of non-linear ballots; see Exercise 4.

3.3 The Equivalence of Arrow's Theorem and the Gibbard–Satterthwaite Theorem

In geometry, we say that two versions of the parallel postulate are equivalent if each becomes a theorem when the other is added as an axiom to Euclid's original four. Similarly, we say that two versions of the axiom of choice are equivalent if each becomes a theorem when the other is added as an axiom to the Zermello–Frankel axioms for set theory.

The reasons these assertions have formal content is that the results whose equivalence is being claimed are independent of the remaining axioms (assuming the consistency of the remaining axioms). Absent this condition of independence, the theorem asserting that $2 + 2 = 4$ would qualify as being equivalent to Andrew Wiles' elliptic curve result that settled Fermat's last theorem (each being provable from the standard axioms of set theory with the other added – or not added, as it turns out).

Equivalence, however, is also used in an informal sense inspired by the formal notion above. We say that two theorems are equivalent if each is "easily derivable" from the other, where the ease of the derivation is measured (intuitively) relative to the difficulty of the stand-alone proofs of the theorems whose equivalence is being asserted. It is in this informal sense that we want to ask about the equivalence of Arrow's theorem and the Gibbard–Satterthwaite theorem.

Initially, we work in the context of linear ballots, and we begin with the original version of Arrow's theorem that took place in the context of social welfare functions (or social choice functions that satisfy transitive rationality, but these naturally correspond to social welfare functions). In what ways do the two theorems differ?

(1) Arrow's theorem pertains to social welfare functions; the Gibbard–Satterthwaite theorem pertains to voting rules.
(2) The Gibbard–Satterthwaite theorem requires resoluteness as an assumption; Arrow's theorem does not.
(3) Arrow's theorem requires Pareto as an assumption (and fails in the presence of non-imposition as shown by an antidictatorship); the Gibbard–Satterthwaite theorem requires only non-imposition.[15]
(4) Non-manipulability implies monotonicity: IIA does not (as shown again by an antidictatorship).

Issues 1 and 2 have been addressed in the exercises at the end of Chapter 1. We showed there that the voting rule version of Arrow's theorem is equivalent to the social welfare version of Arrow's theorem (and this is a very good model for the use of the word "equivalent"). Moreover, we outlined there a direct proof that a voting rule satisfying Pareto and IIA is resolute.

Issues 3 and 4 seem more delicate, and we deal with them by slightly weakening the statement of each theorem as follows:

(a) We replace "non-imposition" by "Pareto" in the Gibbard–Satterthwaite theorem.

[15] There are versions of Arrow's theorem, dating back to Wilson (1972), that do not assume Pareto. For an explicit treatment where Pareto is replaced by non-imposition, see Saari (1995, p. 87).

(b) We replace "IIA" by "MIIA" in Arrow's theorem, where MIIA is a version of independence of irrelevant alternatives that builds in monotonicity for resolute procedures.

More precisely, a resolute voting rule satisfies MIIA if, whenever **P** and **P**′ are profiles and a and b are distinct alternatives such that $\mathrm{V}(\mathbf{P}) = a$ and $\mathrm{V}(\mathbf{P}') = b$, there is some i such that $a\mathrm{P}_i b$ and $b\mathrm{P}'_i a$. With this, we can now prove the following.

Theorem 3.3.1. *For every set* A *of three or more alternatives and every* $\mathrm{n} \geq 1$, *a voting rule for* (A, n) *that satisfies Pareto is resolute and non-manipulable iff it satisfies MIIA.*

Proof: The conditions on either side of the "iff" hold precisely when the procedure is a dictatorship. But this observation misses the point of what we are trying to establish, and so we proceed with direct arguments of the two implications.

Assume first that V is resolute, non-manipulable, and satisfies Pareto. Suppose that **P** and **P**′ are profiles and a and b are distinct alternatives such that $\mathrm{V}(\mathbf{P}) = a$ and $\mathrm{V}(\mathbf{P}') = b$. We want to show that there is some i such that $a\mathrm{P}_i b$ and $b\mathrm{P}'_i a$. Because V is non-manipulable and resolute, we know that it satisfies down-monotonicity by Lemma 3.1.5 (which was short and completely self-contained).

Thus, for each voter j with $b\mathrm{P}_j a$ and $a\mathrm{P}'_j b$, we can plunge b to the bottom of that ballot in P. Moreover, if the alternatives other than a and b are $c_1, \ldots,$ c_k, we can now, for each ballot in **P** and in **P**′, plunge these other alternatives to the bottom of the ballots in the order c_1, then c_2, etc.

Retaining the names **P** and **P**′ for these now-altered profiles, we have that a and b occupy the top two spots on all ballots in both profiles, all other alternatives occur in the order $c_1, \ldots, c_k$ on every ballot, any ballot that has b over a in **P** has b over a in **P**′ (lest b would have been plunged in **P**), and $\mathrm{V}(\mathbf{P}) = a$ and $\mathrm{V}(\mathbf{P}') = b$. Because $\mathbf{P} \neq \mathbf{P}'$, we can choose i such that $\mathrm{P}_i \neq \mathrm{P}'_i$. But this means $a\mathrm{P}_i b$ and $b\mathrm{P}'_i a$, as desired.

Now assume that V satisfies MIIA and Pareto. By Exercise 12 in Chapter 1 (a short and self-contained exercise), we know that V is resolute, so it suffices to show that V is non-manipulable. So suppose that **P** and **P**′ are profiles, i is a voter, $\mathbf{P}|\mathrm{N} - \{i\} = \mathbf{P}'|\mathrm{N} - \{i\}$, $\mathrm{V}(\mathbf{P}) = a$ and $b\mathrm{P}_i a$. Then MIIA implies that $\mathrm{V}(\mathbf{P}') \neq b$ because every voter who has a over b in **P** still has a over b in **P**′. Hence, voter i's attempt at manipulation failed. □

If Theorem 3.3.1 convinces the reader that Arrow's theorem is equivalent to the Gibbard–Satterthwaite theorem, then it's unlikely that other results, old or new, will change the reader's mind. On the other hand, if the reader is left unconvinced of the equivalence, then he or she should not conclude that they fail to be equivalent, but simply that our Theorem 3.3.1 was inadequate for the task at hand. The efforts of others (for example, Bernard Monjardet's discussion in Monjardet, 1999), might be more convincing.

We offer two additional remarks here. First, both Arrow and Gibbard and Satterthwaite proved their theorems directly in the context of ballots that allowed ties. Later, as we said, it was realized that the Gibbard–Satterthwaite theorem in this context could easily be derived from the linear-ballot context (as we did in Section 3.2). It turns out that the same is true of Arrow's theorem – see Exercises 5 and 6. Second, in the case of infinitely many voters, the equivalence of Arrow's theorem and the Gibbard–Satterthwaite theorem evaporates unless one replaces individual manipulability with coalitional manipulability (Section 6.2) and, even there, differences arise if there are infinitely many alternatives as well as infinitely many voters (Section 6.4).

3.4 Reflections on the Proof of the Gibbard–Satterthwaite Theorem

A couple of the pieces of the proof of the Gibbard–Satterthwaite theorem are worth isolating. For example, Lemma 3.1.9 actually holds for every binary relation in the following sense (see Blau, 1972, Monjardet, 1978, and Makinson, 1996).

Proposition 3.4.1 (The All-or-None Lemma). *Suppose that A has three or more elements and that* ß *is an irreflexive binary relation on A that satisfies the following: for any three distinct alternatives a, b, c* $\in$ *A,*

(1) if aßb, *then* aßc, *and*
(2) if aßb, *then* cßb.

Then either ß $= \emptyset$ *or* ß $= A \times A - \Delta$, *where* $\Delta = \{(a, a) : a \in A\}$.

Proof: Assume that ß $\neq \emptyset$, and choose $a, b \in$ A such that aßb. Now let $(x, y) \in$ A x A $- \Delta$. We show that $(x, y) \in$ ß.

Case 1. $y \neq a$
Because aßb, we know by (1) that aßy. Because $x \neq y$, we can now apply (2) to get xßy and so $(x, y) \in$ ß.
Case 2. $x \neq b$
Because aßb, we know by (2) that xßb. Because $y \neq x$, we can now apply (1) to get xßy and so $(x, y) \in$ ß again.
Case 3. $y = a$ and $x = b$

Because A has three or more elements, we can choose c distinct from a and b. Now, because $aßb$, we know by (1) that $aßc$, and by (2) that $bßc$. A final application of (1) shows that $bßa$, and so $xßy$. Thus, $(x, y) \in ß$, as desired. This completes the proof. □

The second piece of the proof of the Gibbard–Satterthwaite theorem that is worth isolating for future reference is somewhat more voting theoretic, although there is no voting rule explicitly involved in the following proposition.

Proposition 3.4.2. *Suppose that for each set X $\subseteq$ N we have an antireflexive binary relation, also denoted by X, on the set A of three or more alternatives. Say that a set X $\subseteq$ N is a dictating set if* aXb *for every distinct pair* a *and* b *of alternatives in A, and assume that the set N itself is a dictating set and that* a∅b *fails for every* a *and* b*. Suppose the following holds:*

The splitting condition: If aXb*, and* c *is distinct from* a *and* b*, then for every partition of X into disjoint subsets Y and Z (one of which may be empty), either* a*Y*c *or* cZb.

Then for every dictating set X there exists an i $\in$ *X such that* {i} *is also a dictating set.*

Proof: We first claim that if aXb for some pair a and b of alternatives, then X is a dictating set. To see this, we respectively let Y = ∅ and then Z = ∅ in the splitting condition. Because we never have a∅b for any a and b, the assumptions in Proposition 3.4.1 are satisfied by the binary relation X. Thus, either aXb fails for every a and b, which we are assuming is not the case, or X is a dictating set, as desired.

Now, if X is a dictating set, then we can choose Y $\subseteq$ X to be a non-empty dictating set of minimal size among all subsets of X. If Y were not a singleton, then we could once again appeal to the splitting condition to arrive at a proper subset of Y that is, by the first paragraph of this proof, again a dictating set, thus contradicting the minimality of Y. □

In the proof of the Gibbard–Satterthwaite theorem, the relation X was defined in terms of a resolute voting rule V by asserting that aXb holds provided that, for every profile **P** in which every voter in X ranks a over b on his or her ballot, we have V(**P**) $\neq$ b. But this is only one such possibility. And, although we later make use of Proposition 3.4.2 in the context of non-manipulability, let us here illustrate its use with two different interpretations of "aXb" in the context of Arrow's theorem. In each case, we will assume a version of Pareto and a version of IIA, and then we will prove an "existence lemma" and a "splitting lemma." Proposition 3.4.2 will then immediately give as some kind of dictatorial behavior.[16]

For simplicity, we assume in the next three theorems that we have three or more alternatives, linear ballots, and that the aggregation procedures are all monotone. Extensions that eliminate these latter two assumptions are left to the reader (see Exercises 7 and 8), but such extensions then yield proofs of the results stated in Chapter 1 as Theorems 1.3.1, 1.3.2, and 1.3.3.

Theorem 3.4.3. *Suppose that V is a monotone voting rule in the context of linear ballots and three or more alternatives, and that V satisfies the following:*

(1) Pareto: if everyone ranks a *over* b*, then* b *is a non-winner.*
(2) CIIA: if a *is a winner and* b *is a non-winner, and ballots are changed, but everyone who had* a *over* b *keeps* a *over* b *and vice-versa, then, in the new election,* b *is still a non-winner.*

Then V is resolute and there is a dictator.

Proof: For X $\subseteq$ N and $a, b \in$ A, say that aXb if, for every profile **P** in which everyone in X has a over b on their ballots, $b \notin$ V(**P**).

Claim 1 (the existence lemma): Suppose there exists a profile **P** in which everyone in X has a over b, everyone not in X has b over a, and for which $a \in$ V(**P**) and $b \notin$ V(**P**). Then aXb.

Proof. Suppose, for contradiction, that aXb fails. Then we have a profile **P**′ in which everyone in X has a over b, and yet $b \in$ V(**P**′). By monotonicty of V we can assume, in **P**′, that everyone not in X has b over a. But now we can change the ballots in **P** so that they become identical to those in **P**′ and, by CIIA, conclude that b is still a non-winner. This contradiction completes the proof of Claim 1.

Claim 2 (the splitting lemma): If aXb, and c is distinct from a and b, then for every partition of X into disjoint sets Y and Z (one of which may be empty), either aYc or cZb.

Proof. Consider the following profile P, in which every voter places all alternatives other than a, b, and c below these three in any order

[16] For a series of different proofs of Arrow's theorem, the reader can start with Fishburn (1970), Blau (1972), Barberá (1980), and Geanakoplos (1996).

whatsoever:

Y	Z	N–X
a	c	b
b	a	c
c	b	a

By Pareto, the winners are among a, b, and c. Because $a\text{X}b$, b is a non-winner. If c is a winner, then $c\text{Z}b$ by the existence lemma. If c is a non-winner, then a must be a winner (being the only alternative not yet ruled out). Thus, $a\text{Y}c$ by the existence lemma, and this proves Claim 2.

Notice that Pareto implies that $a\text{N}b$ for every a and b, but that it, together with CIIA, implies $a\emptyset b$ fails for every a and b (because, in particular, if everyone has b on top of his or her ballot, then b is the unique winner). It now follows from Proposition 3.4.2 that there is a voter i such that $a\{i\}b$ for every a and b. But this clearly means that the election winner is always the unique alternative that voter i has on top of his or her ballot. This completes the proof of Theorem 3.4.3.

□

The version of monotonicity that we need for the next result, dealing with social welfare functions, asserts that if a is not over b on the final list, and some voter who had a over b on his or her ballot interchanges a and b, then a is still not over b on the final list.

Theorem 3.4.4. *Suppose that V is a monotone social welfare function in the context of linear ballots and three or more alternatives, and that V satisfies the following:*

(1) Pareto: if everyone ranks a *over* b, *then* a *is over* b *on the final list.*
(2) IIA: if a *is over* b *on the final list, and ballots are changed, but everyone who had* a *over* b *keeps* a *over* b *and vice versa, then, in the new election,* a *is still over* b *on the final list.*

Then V is resolute and there is a dictator.

Proof: For X $\subseteq$ N and a, $b \in$ A, say that $a\text{X}b$ if a is over b on the final list whenever everyone in X has a over b on their ballots.

Claim 1 (the existence lemma): Suppose there exists a profile **P** in which everyone in X has a over b, everyone not in X has b over a, and for which a is over b on the final list V(**P**). Then $a\text{X}b$.

Proof. Suppose, for contradiction, that aXb fails. Then we have a profile **P**′ in which everyone in X has a over b, and yet a is not over b on the final list V(**P**′). By monotonicty of V, we can assume that everyone not in X has b over a in **P**. But now we can change the ballots in **P** so that they become identical to those in **P**′ and, by CIIA, conclude that a is still over b on the final list. This contradiction completes the proof of Claim 1.

Claim 2 (the splitting lemma): If aXb, and c is distinct from a and b, then for every partition of X into Y ∪ Z, either aYc or cZb.

Proof. Consider the following profile **P**, where every voter places all alternatives other than a, b, and c below these three in any order whatsoever:

Y	Z	N–X
a	c	b
b	a	c
c	b	a

By Pareto, a, b, and c are ranked above all other alternatives in the final list. Because aXb, a is over b on the final list. If c is over b on the final list, then cZb by the existence lemma. If c is not over b on the final list, then a is over c on the final list (because a is over b, and b is over c or tied with c on the final list). Thus, aYc by the existence lemma, and this proves Claim 2.

Notice that Pareto not only implies that aNb for every a and b, but that $a\emptyset b$ fails for every a and b. It now follows from Proposition 3.4.2 that there is a voter i such that $a\{i\}b$ holds for every a and b. But this clearly means that the final list is identical to voter i's ballot. This completes the proof of Theorem 3.4.3. □

Because we know what monotonicity means in the context of a social welfare function, there is no need to define what it means in the context of a social choice function that satisfies transitive rationality.

Theorem 3.4.5. *Suppose that V is a monotone social choice function satisfying transitive ratonality in the context of linear ballots and three or more alternatives, and that V satisfies the following:*

(1) Pareto: if everyone ranks a *over* b *then* a *is over* b *on the final list.*
(2) IIA: if a *is over* b *on the final list, and ballots are changed, but everyone who had* a *over* b *keeps* a *over* b *and vice versa, then, in the new election,* a *is still over* b *on the final list.*

Then V is resolute and there is a dictator.

Proof: Because we are assuming transitive rationality, the social choice function V is really a social welfare function. Hence, Theorem 3.4.5 follows immediately from Theorem 3.4.4. □

3.5 Exercises

(1) [S] Suppose that N is an eight-element set and that G is a collection of subsets of N such that (i) N $\in$ G, and (ii) for every X $\subseteq$ N, if X $\in$ G and X is partitioned into sets Y and Z, then either Y $\in$ G or Z $\in$ G. Give three different proofs that $\{i\} \in$ G for some $i \in$ N.

(2) [S] Prove that a resolute voting rule that satisfies down-montonicity also satisfies monotonicity.

(3) [S] Prove that in the context of non-linear ballots and three or more alternatives, there are weak dictatorships that are manipulable.

(4) [S] Prove that in the context of non-linear ballots and three or more alternatives, there are weak dictatorships that are non-manipulable.

The following two exercises allow one to derive a version of Arrow's theorem for non-linear ballots from a version for linear ballots. Notice, however, that we are using MIIA (the version of IIA that incorporates monotonicity).

(5) [T] Suppose that V satisfies MIIA and that for every linear ballot **P**, $\mathrm{V}(\mathbf{P}) = \{\mathrm{top}_i(\mathbf{P})\}$. Prove that for every ballot $\mathbf{P}'$, $\mathrm{top}_i(\mathbf{P}') \in \mathrm{V}(\mathbf{P}')$.

(6) Derive Arrow's theorem for non-linear ballots from the version for linear ballots.

(7) [T] Show that the assumption of monotonicity can be eliminated from Theorem 3.4.5 by establishing that if $b \notin \mathrm{V}(\mathbf{P})$ for every profile **P** in which everyone in X has a over b on their ballots, and everyone else has b over a on their ballots, then $a\mathrm{X}b$.

(8) [T] Do a version of Exercise 7 that eliminates the monotonicity assumption from Theorem 3.4.4 and Theorem 3.4.5.

(9) [S not T] Prove that if V satisfies Pareto and can be manipulated by a coalition, then it can be manipulated by a single voter.

PART TWO

4
Non-Resolute Voting Rules

4.1 The Duggan–Schwartz Theorem

We begin by restating, with a slight extension for non-linear ballots, the definition of manipulability that we use in the first two sections of this chapter.

Definition 4.1.1. In the context of linear or non-linear ballots, a voting rule V can be *manipulated by an optimistic voter* if there exists a profile $\mathbf{P} = (\mathrm{R}_1, \ldots, \mathrm{R}_n)$, which we think of as giving the true preferences of the n voters, and another ballot Q_i, which we think of as a disingenuous ballot from voter i such that, letting $\mathbf{P}' = (\mathrm{R}_1, \ldots, \mathrm{R}_{i-1}, \mathrm{Q}_i, \mathrm{R}_{i+1}, \ldots, \mathrm{R}_n)$,

$$\exists x \in \mathrm{V}(\mathbf{P}') \text{ such that } \forall y \in \mathrm{V}(\mathbf{P}),\ x\mathrm{P}_i y.$$

Similarly, a voting rule V can be *manipulated by a pessimistic voter* if there exists a profile $\mathbf{P} = (\mathrm{R}_1, \ldots, \mathrm{R}_n)$, which we think of as giving the true preferences of the n voters, and another ballot Q_i, which we think of as a disingenuous ballot from voter i such that, letting $\mathbf{P}' = (\mathrm{R}_1, \ldots, \mathrm{R}_{i-1}, \mathrm{Q}_i, \mathrm{R}_{i+1}, \ldots, \mathrm{R}_n)$,

$$\forall x \in \mathrm{V}(\mathbf{P}')\ \exists y \in \mathrm{V}(\mathbf{P}) \text{ such that } x\mathrm{P}_i y.$$

More briefly, a voting rule can be manipulated by an optimist if there is at least one election in which some voter can submit a disingenuous ballot and improve the max of the set of winners according to his or her true preferences; i.e., at least one of the winners from $\mathbf{P}' = (\mathrm{R}_1, \ldots, \mathrm{R}_{i-1}, \mathrm{Q}_i, \mathrm{R}_{i+1}, \ldots, \mathrm{R}_n)$ is – according to R_i – strictly preferred to all of the winners from $\mathbf{P} = (\mathrm{R}_1, \ldots, \mathrm{R}_n)$.

Similarly, a voting rule can be manipulated by a pessimist if there is at least one election in which some voter can submit a disingenuous ballot and improve the min of the set of winners according to his or her true preferences; i.e., all of the winners from $\mathbf{P}' = (\mathrm{R}_1, \ldots, \mathrm{R}_{i-1}, \mathrm{Q}_i, \mathrm{R}_{i+1}, \ldots, \mathrm{R}_n)$ are – according to R_i – strictly preferred to at least one of the winners from $\mathbf{P} = (\mathrm{R}_1, \ldots, \mathrm{R}_n)$.

Recall that a voting rule V is said to be non-imposed if, for every alternative x, there exists at least one profile **P** such that V(**P**) = $\{x\}$. We also said that, in the context of linear ballots, a voter would be called a nominator for a voting rule V if, for every profile **P** and every alternative x, if that voter has x on top of his or her ballot, then $x \in$ V(**P**).

With these preliminaries, we can turn to a generalization of the Gibbard–Satterthwaite theorem that handles ties in the outcome of elections (although we are still disallowing ties in the ballots for the moment). It was first established by John Duggan and Thomas Schwartz in a 1993 preprint entitled "Strategic manipulability is inescapable: Gibbard–Satterthwaite without resoluteness." It was never published in exactly that form, but a weaker version appeared in Duggan and Schwartz (2000). The proof we give here is from Taylor (2002) and is quite distinct from either of those; it follows the outline that we used in proving the Gibbard–Satterthwaite theorem. For other significant contributions along these same lines, see Barberá, Dutta, and Sen (2001), Campbell and Kelly (2000 and 2002), Ching and Zhou (2002), and, for an early precursor, Feldman (1979b).

Theorem 4.1.2 (The Duggan–Schwartz Theorem for Linear Ballots). *In the context of linear ballots, if* n *is a positive integer and* A *is a set of three or more alternatives, then any voting rule for* (A, n) *that is non-imposed and cannot be manipulated by an optimist or a pessimist has a nominator.*

Our starting point in the proof is with a slightly strengthened version of one of Duggan and Schwartz's observations. First, a piece of terminology: If **P** is a profile, then a set X of alternatives is said to be a *top set* (for **P**) if each voter prefers (according to his or her ballot) every alternative in X to every alternative not in X. For example, if every voter has x on top of his or her ballot, then $\{x\}$ is a top set.

Suppose now that V is a voting rule that cannot be manipulated by an optimist or a pessimist, and that **P** is a profile for which X is a top set. Assume V is non-imposed (although it suffices to assume that there is at least one profile **P**′ with V(**P**′) $\subseteq$ X). Then we claim that V(**P**) $\subseteq$ X. If not, we could convert **P** to **P**′, one ballot at a time, until the set of winners first changes from not being a subset of X, which we are assuming is true with **P**, to being a subset of X, which we are assuming is true with **P**′. If this occurs as we change ballot B_i to C_i, then we can take B_i to be the true preferences of voter i and see that his or her insincere submission of C_i has improved the min from something not in X to something in X. This proves the claim.

Now, mimicking what we did with the proof of the Gibbard–Satterthwaite theorem in Chapter 3, we introduce a definition that applies to some fixed voting rule V.

Definition 4.1.3. If X is a set of voters, and a and b are two distinct alternatives in the set A, then *X can use a* to block b, denoted $a\mathrm{X}b$, if, for every profile **P** in which all the voters in X rank a over b on their ballots, $\mathrm{V}(\mathbf{P}) \neq \{b\}$. The set X is a *dictating set* if $a\mathrm{X}b$ for every distinct pair a, b of alternatives in A.

Definition 4.1.4. A voting rule V satisfies *down-monotonicity for singleton winners* provided that, for every profile **P** with $|\mathrm{V}(\mathbf{P})| = 1$, if $\mathbf{P}'$ is the profile obtained from **P** by having one voter move one losing alternative down one spot on his or her ballot, then $\mathrm{V}(\mathbf{P}') = \mathrm{V}(\mathbf{P})$.

Notice that if V satisfies down-monotonicity for singleton winners and $\mathrm{V}(\mathbf{P}) = \{x\}$, then $\mathrm{V}(\mathbf{P}') = \{x\}$ whenever $\mathbf{P}'$ is derived from **P** by having *several* voters move *several* alternatives down *several* slots on their ballots. As with the Gibbard–Satterthwaite theorem, this observation will be important in applying down-monotonicity.

For an example of a voting rule that satisfies down-monotonicity for singleton winners, consider again the omninomination rule from Chapter 2. Here, the only way that we have a single winner in an election is if every voter has this same alternative at the top of his or her ballot. But, in this case, the lowering of any losing alternative by any voter has no effect on the outcome.

The plurality rule, on the other hand, is an example of a voting rule that does not satisfy down-montonicity for singleton winners. For example, consider the following profile **P**:

P

a	a	b	c
b	b	a	b
c	c	c	a

With the plurality rule, the winner is $\{a\}$, but if the voter on the far right moves the losing alternative c down one space (interchanging c and b), then a and b are tied for the win. Notice that this also is an example of manipulation by an optimistic voter because the max for this voter has gone from a to b.

Lemma 4.1.5. *Every voting rule that cannot be manipulated by an optimist or a pessimist satisfies down-monotonicity for singleton winners.*

Proof: Assume that down-monotonicity for singleton winners fails. Then there exist two profiles **P** and $\mathbf{P}'$ and an alternative y such that:

(1) In **P**, voter i ranks y directly over x, $\mathrm{V}(\mathbf{P}) = \{w\}$, and $w \neq y$ (that is, y is the losing alternative that voter i will be moving down).

(2) $\mathbf{P}'$ differs from $\mathbf{P}$ only in that voter i has interchanged the position of x and y on his or her ballot, and yet $v \in \mathrm{V}(\mathbf{P}')$ for some $v \neq w$.
Assuming (1) and (2), we will show that the system can, in fact, be manipulated.

Case 1: v is over w on voter i's ballot in $\mathbf{P}$.
In this case, we can regard voter i's ballot in $\mathbf{P}$ as giving his or her true preferences. Thus, if voter i submits his or her sincere ballot $\mathbf{P}$, w is the only winner, although voter i prefers v to w. But if he or she submits a disingenuous ballot (the one in $\mathbf{P}'$), then he or she improves the max from w to at least v.
Case 2: w is over v on voter i's ballot in $\mathbf{P}'$.
In this case, we can regard voter i's ballot in $\mathbf{P}'$ as giving his or her true preferences. Thus, if voter i submits his or her sincere ballot $\mathbf{P}'$, v is among the winners, although voter i prefers w to v. But if he or she submits a disingenuous ballot (the one in $\mathbf{P}$), then w is, in fact, the only winner, and so he or she improves the min from v or worse to w.
Case 3: Otherwise.
In this case, w is over v on voter i's ballot in $\mathbf{P}$, and v is over w on voter i's ballot in $\mathbf{P}'$. But this means that we have $w = y$ and $v = x$, contradicting our assumption that $y \neq w$. □

At this point, we are ready to set up for our appeal to Proposition 3.4.2. For the remainder of the proof, we assume that V is a voting rule for (A, n) in the context of linear ballots, and that V is non-imposed and cannot be manipulated by an optimist or a pessimist, and therefore satisfies down-monotonicity for singleton winners.

Lemma 4.1.6 (The Existence Lemma). *Suppose that X is a set of voters and that* a *and* b *are two distinct alternatives in A. Then, in order to show that* aXb, *it suffices to produce one profile* ***P*** *for which:*

(ii) {a,b} *is a top set,*
(iii) everyone in X ranks a *over* b,
(iv) everyone else ranks b *over* a, *and*
(v) a ∈ *V(**P**).*

Proof: Suppose we have such a profile $\mathbf{P}$ but $a\mathrm{X}b$ fails. Then we also have a profile $\mathbf{P}'$ in which everyone in X ranks a over b and $\mathrm{V}(\mathbf{P}') = \{b\}$. Using down-monotonicity for singleton winners, we can convert $\mathbf{P}'$ into the profile $\mathbf{P}$ assumed to exist, and get $\mathrm{V}(\mathbf{P}) = \{b\}$. But this is a contradiction because $a \in \mathrm{V}(\mathbf{P})$. □

Lemma 4.1.7 (The Splitting Lemma). *Suppose X is a set of voters and that* a, b *and* c *are distinct alternatives in A. Assume also that* a*X*b, *and that X is partitioned into disjoint sets Y and Z (one of which may be empty). Then either* a*Y*c *or* c*Z*b.

Proof: Consider any profile **P** in which every voter in Y has a first, b second, and c third; every voter in Z has c first, a second, and b third; and everyone else (i.e., those in N−X) has b first, c second, and a third. We can picture these ballots as follows:

Y	Z	N−X
a	c	b
b	a	c
c	b	a
.	.	.
.	.	.
.	.	.

Because $\{a, b, c\}$ is a top set, our previous discussion guarantees that V(**P**) $\subseteq$ $\{a, b, c\}$. Because aXb, V(**P**) $\neq \{b\}$, and so either $a \in$ V(**P**) or $c \in$ V(**P**).

Case 1: $a \in$ V(**P**).

For each voter in Y, we one by one move b just below c. As we do this – changing a ballot from B_i to C_i – a remains a winner, or else we could regard C_i as the true preferences and then have voter i improving his or her max from something other than his or her top choice to his top choice a. Now, for every voter not in Y or Z (i.e., those in N–X) we one by one move b just below a. Again, as we do this – changing a ballot from B_i to C_i – a remains a winner, or else we could regard B_i as the true preferences and then have voter i improving his or her min from a to b or c. But now we have produced a profile **P**′ in which $\{a, c\}$ is a top set, everyone in Y prefers a to c, everyone else prefers c to a, and in which $a \in$ V(**P**′). Thus, by Lemma 4.1.6 (the existence lemma), we have aYc as desired.

Case 2: $c \in$ V(**P**).

For each voter in Z, we one by one move a just below b. As we do this – changing a ballot from B_i to C_i – c remains a winner (or else we could regard C_i as the true preferences and then have voter i improving his or her max from something other than his or her top choice to his or her top choice c). Now, for every voter in Y, we one by one move a just below c. Again, as we do this – changing a ballot from B_i to C_i – c remains a winner (or else we could regard B_i as the true preferences and then have voter i improving his or her min from c to a or b). But now we have produced a profile $\mathbf{P}'$ in which $\{b, c\}$ is a top set, everyone in Z prefers c to b, everyone else prefers b to c, and in which $c \in \mathrm{V}(\mathbf{P}')$. Thus, by Lemma 4.1.6, we have $c\mathrm{Z}b$ as desired. □

Lemma 4.1.8 *If X is a dictating set, then there exists a voter* i $\in$ *X such that* {i} *is a dictating set. Thus, voters* i*'s top choice is the unique winner whenever the winner is a singleton.*

Proof: The result follows immediately from Proposition 3.4.2 and Lemma 4.1.7, once we establish that for every pair a and b of alternatives, $a\mathrm{N}b$ holds and $a\emptyset b$ fails.

For the former, suppose that $a\mathrm{N}b$ fails, and choose a profile $\mathbf{P}$ in which every voter has a over b on his or her ballot, but $\mathrm{V}(\mathbf{P}) = \{b\}$. Now choose a profile $\mathbf{P}'$ such that $\mathrm{V}(\mathbf{P}') = \{a\}$. Using down-monotonicity for singleton winners, we can first move b to the bottom of every ballot in $\mathbf{P}'$ and then move each of the other losing alternatives below b in some fixed order. Similarly, we can move all alternatives other than a and b to the bottom, in this same fixed order, of all the ballots in $\mathbf{P}$. But then we have identical profiles with two different election outcomes.[17]

For the latter, we can see that $a\emptyset b$ fails simply by choosing a profile for which $\mathrm{V}(\mathbf{P}) = \{b\}$. Then every voter in $\emptyset$ (there are none) has a over b, but $\mathrm{V}(\mathbf{P}) = \{b\}$. This completes the proof of Lemma 4.1.8. □

To complete the proof of Theorem 4.1.2, we need only one additional observation, still assuming that V is a voting rule that is non-imposed and cannot be manipulated by an optimist or a pessimist. Suppose voter i's top choice is the unique winner whenever the winner is a singleton (as guaranteed by Lemma 4.1.8). Then, we claim that voter i's top choice is always among the set of winners.

The argument here runs as follows. Suppose not, and choose a profile $\mathbf{P}$ such that the $x \notin \mathrm{V}(\mathbf{P})$ where x is the alternative that is on top of voter i's ballot and such that $|\mathrm{V}(\mathbf{P})|$ is as small as possible. We can't have $|\mathrm{V}(\mathbf{P})| = 1$ by our assumption that voter i's top choice is the unique winner whenever the winner is a singleton.

[17] We can reach this same conclusion with "$\forall x \exists \mathbf{P}$ with $\mathrm{V}(\mathbf{P}) = \{x\}$" replaced by the weaker assumption "$\forall x \exists \mathbf{P}$ with $x \in \mathrm{V}(\mathbf{P})$" if we replace down-monotonicity with the direct assumption that the voting rule cannot be manipulated by an optimist or a pessimist. That is, if every voter has a over b and $\mathrm{V}(\mathbf{P}) = \{b\}$, then we can use down-monotonicity to make $\{a, b\}$ a top set. Now choose $\mathbf{P}'$ such that $a \in \mathrm{V}(\mathbf{P}')$. Convert $\mathbf{P}$ to $\mathbf{P}'$, one ballot at a time, until a becomes a winner. At this point, the voter who just changed his or her ballot has improved his or her max to his most preferred alternative a.

Assume V(**P**) = $\{s_1, \ldots, s_t\}$ with $t \geq 2$ and $x \notin$ V(**P**), and assume voter i ranks s_1 over s_2 over . . . over s_t. Let **P**′ be any profile in which voter i's ballot is the same as in **P**, but in which all the other voters have $s_1, \ldots, s_t$ as a top set in that order. Now change **P** to **P**′ one ballot at a time.

We first claim that as we change a ballot from B_j to C_j, no new alternative w gets added to the set V(**P**) of winners, because we could then regard C_j as the true preferences of that voter, and the disingenuous submission of B_j would then improve the min from w or worse to s_t. Notice that this argument covers $x = w$ as well.

Moreover, no s_i can be lost from V(**P**) by the minimality of |V(**P**)| – this is why we needed to observe that x is not added to V(**P**). But now, starting with **P**′, voter i can bring s_1 to the top of his or her ballot and make the set of winners a singleton $\{s_1\}$ (because $\{s_1\}$ is then a top set), thus improving his or her minimum because $t \geq 2$. This completes the proof of Theorem 4.1.2. □

4.2 Ties in the Ballots

We showed in Section 3.2 that in the Gibbard–Satterthwaite context, if ties are allowed in the ballots (but not in the outcome), and if V is non-imposed, then there is dictator in the sense that the winner must be one of the alternatives tied for top on his or her ballot. In the Duggan–Schwartz context, we can do a similar thing.

Theorem 4.2.1 (The Duggan–Schwartz Theorem for Non-linear Ballots). *In the context of non-linear ballots, if* n *is a positive integer and A is a set of three or more alternatives, then any voting rule V for* (A, n) *that is non-imposed and cannot be manipulated by an optimist or a pessimist is a dictatorship in the sense that then there is a voter such that, for every profile **P**, the set V(**P**) contains at least one of the alternatives that this voter has tied for top.*

Proof: If V is the voting rule postulated, let V′ be the restriction of V to linear profiles. Clearly V′ also cannot be manipulated by either an optimist or a pessimist. Moreover, we claim that if **P** is a profile in which every voter has x on top, then V(**P**) = $\{x\}$. To see this, choose **P**′ (consisting of ballots that might have ties) such that V(**P**′) = $\{x\}$. Change **P** to **P**′ one ballot at a time until the winner becomes $\{x\}$. At this point some voter has improved the min to be his or her most preferred alternative x.

It now follows from Theorem 4.1.2 that if all the ballots are linear, then the top alternative on voter i's ballot is among the winners. Assume, for contradiction, that **P** is a profile (with ties) such that no alternative from voter i's top block is among the winners, and choose such a profile $\mathbf{P}$ so that $|V(\mathbf{P})|$ is as small as possible. One ballot at a time, move the set $V(\mathbf{P})$ to the top of everyone's (except voter i's) ballot and break all ties in these ballots. No new winner w is added as we do this, or we could regard the ballots with $V(\mathbf{P})$ on top (and no ties) as the true preferences, and see that the min has been improved from w to something in $V(\mathbf{P})$. Moreover, nothing in $V(\mathbf{P})$ is lost by the minimality of $|V(\mathbf{P})|$.

Finally, we can break all the ties in voter i's ballot, in which case some alternative that was in his or her top block becomes one of the winners. At this point, voter i has improved his or her max from something not in the top block to something in the top block. This completes the proof of Theorem 4.2.1. □

4.3 Feldman's Theorem

Recall that a voting rule is expected-utility manipulable if there exist profiles $\mathbf{P}$ and $\mathbf{P}'$ and a voter i such that $\mathbf{P} \mid N - \{i\} = \mathbf{P}' \mid N - \{i\}$ and voter i, whose true preferences we take to be given by his or her ballot in $\mathbf{P}$, prefers the election outcome X from $\mathbf{P}'$ to the election outcome Y from $\mathbf{P}$ in the following sense: There exists a utility function u representing P_i such that, if $p(x) = 1/|X|$ for every $x \in X$, and $p(y) = 1/|Y|$ for every $y \in Y$, then $\sum\{p(x) \cdot u(x): x \in X\} > \sum\{p(x) \cdot u(y): y \in Y\}$.

If u is a utility function representing P_i and $p = (p_1, p_2)$ is a fixed pair of probability functions, with p_1 on X and p_2 on Y, then we let $E_{u,p,i}(X)$ and $E_{u,p,i}(Y)$ denote the corresponding expected utilities. That is,

$$E_{u,p,i}(X) = \sum\{p_1(x) \cdot u(x) : x \in X\}$$

and

$$E_{u,p,i}(Y) = \sum\{p_2(y) \cdot u(y) : y \in Y\}.$$

Our starting point with Feldman's theorem is the following question: What is the combinatorial equivalent of the existence of a utility function u representing P_i such that $E_{u,p,i}(X) > E_{u,p,i}(Y)$, where $p = (p_1, p_2)$ is the pair of probability functions given by $p_1(x) = 1/|X|$ for every $x \in X$, and $p_2(y) = 1/|Y|$ for every $y \in Y$?

To answer this question, we need to consider the more general version in which the probability functions p_1 and p_2 are arbitrary (as opposed to just $p_1(x) = 1/|X|$ for every $x \in X$, and $p_2(y) = 1/|Y|$ for every $y \in Y$.) The point is that it is the stronger result that seems to carry along the inductive argument.

Stating the combinatorial equivalent of "$\exists u, E_{u,p,i}(X) > E_{u,p,i}(Y)$" requires a piece of notation: If R_i is a weak ordering of the set A and $z \in A$, then we'll let G(z) – "G" for "greater" – be the following set:

$$G(z) = \{x \in A : xR_i z\}.$$

Notice that $z \in G(z)$, even if the ballots are linear. With this notation at hand, we can now characterize the desired relation on sets of alternatives.

Proposition 4.3.1. *Fix a linear ordering R_i of the set A, and let X and Y be any two distinct subsets of A. For notation, let* u *be a variable ranging over utility functions that represent R_i and let* (p_1, p_2) *be a fixed pair of strictly positive probability functions with* p_1 *on X and* p_2 *on Y. Then the following are equivalent:*

(1) $\sum\{p_1(x) : x \in X \cap G(z)\} > \sum\{p_2(y) : y \in Y \cap G(z)\}$ *for some* $z \in A$.
(2) $\exists u,\ E_{u,p,i}(X) > E_{u,p,i}(Y)$.

Proof: Assume (1) holds and choose $z \in A$ such that $r > s$ where

$$\sum\{p_1(x) : x \in X \cap G(z)\} = r$$

and

$$\sum\{p_2(y) : y \in Y \cap G(z)\} = s.$$

The desired u will be constructed by setting $u(a)$ very close to one if $a \in G(z)$ and very close to zero if $a \notin G(z)$. More precisely, choose ε such that $0 < \varepsilon < (r - s)/(r - s + 1)$. Define u such that if $a \in G(z)$, then $1 > u(a) > 1 - \varepsilon$, and if $a \notin G(z)$, then $0 < u(a) < \varepsilon$. Then $E_{u,p,i}(X) > r(1 - \varepsilon)$ and $E_{u,p,i}(Y) < s + \varepsilon(1 - s)$, and it is easy to see that $r(1 - \varepsilon) > s + \varepsilon(1 - s)$ because $(r - s)/(r - s + 1) > \varepsilon$. This proves that (2) holds.

Now assume that (1) fails. Thus, $\forall z \in A \sum\{p_1(x) : x \in X \cap G(z\}) \leq \sum\{p_2(y) : y \in Y \cap G(z\})$. Assume that u is arbitrary. We'll show that $E_{u,p,i}(X) \leq E_{u,p,i}(Y)$, proceeding inductively on $|X \cup Y|$. If $|X \cup Y| = 2$, then the result is trivial because both X and Y are singletons. Assume that $|X \cup Y| > 2$. Suppose first that $Y - X \neq \emptyset$, and let $y_0 = \max(Y - X)$. Notice that we can't have $X \subseteq G(y_0)$, so $(X \cup Y) - G(y_0) \neq \emptyset$. Let $w = \max((X \cup Y) - G(y_0))$. Let $Y' = (Y - \{y_0\}) \cup \{w\}$. Then $|X \cup Y'| < |X \cup Y|$. Define p_2' on Y' by

$$p_2'(y) = \begin{cases} p_2(y) & \text{if } y \neq w \\ p_2(y_0) & \text{if } y = w \text{ and } w \notin Y \\ p_2(y_0) + p_2(w) & \text{if } y = w \text{ and } w \in Y. \end{cases}$$

We claim that $\forall z \in A \sum\{p_1(x) : x \in X \cap G(z)\} \leq \sum\{p_2'(y) : y \in Y' \cap G(z)\}$.

Case 1: $z \leq w$ or $z > y_0$

In this case $\sum\{p_2'(y)\colon y \in Y' \cap G(z)\} = \sum\{p_2(y)\colon y \in Y \cap G(z)\}$.

Case 2: $w < z \leq y_0$

If $y_0 = \max(A)$, then $|X \cap G(z)| = 0$, so the result is immediate. If $y_0 \neq \max(A)$, then let v be the smallest alternative that is larger than y_0. Then $X \cap G(z) = X \cap G(v)$, so

$$\begin{aligned}
&\sum\{p_2'(y)\colon y \in Y' \cap G(z)\} \\
&\geq \sum\{p_2(y)\colon y \in Y \cap G(v)\} \\
&\geq \sum\{p_1(x)\colon x \in X \cap G(v)\} \\
&= \sum\{p_1(x)\colon x \in X \cap G(z)\}.
\end{aligned}$$

Moreover, because $|X \cup Y'| < |X \cup Y|$, our inductive hypothesis yields that $E_{u,p,i}(X) \leq E_{u,p,i}(Y')$. But Y' was obtained from Y by shifting the probability that was on y_0 down to the less preferred alternative w. Hence $E_{u,p,i}(Y') \leq E_{u,p,i}(Y)$, and so $E_{u,p,i}(X) \leq E_{u,p,i}(Y)$, as desired.

Now assume that $Y - X = \emptyset$, so $Y \subseteq X$. Let $x_0 = \max(X - Y)$. Then $\sum\{p_1(x)\colon x \in X \cap G(x_0)\} \geq p(x_0) > 0$, so $\sum\{p_2(y)\colon y \in Y \cap G(x_0)\} > 0$, so there is some $y \in Y$ (and thus $y \in X$ also) with $x_0 < y$. Let $w = \min(\{y \in Y : x_0 < y\})$. Let $X' = X - \{x_0\}$ and note that $w \in X'$. Then $|X' \cup Y| < |X \cup Y|$. Define p_1' on X' by

$$p_1'(x) = \begin{cases} p_1(x) & \text{if } x \neq w \\ p_1(x_0) + p_1(w) & \text{if } x = w \end{cases}$$

We claim that $\forall z \in A \sum\{p_1'(x)\colon x \in X' \cap G(z)\} \leq \sum\{p_2(y)\colon y \in Y \cap G(z)\}$.

Case 1: $z \leq x_0$ or $z > w$.

In this case $\sum\{p_1'(x)\colon x \in X' \cap G(z)\} = \sum\{p_1(x)\colon x \in X \cap G(z)\}$.

Case 2: $x_0 < z \leq w$.

If $x_0 = \max(A)$, then $|X \cap G(z)| = 0$, so the result is immediate. If $x_0 \neq \max(A)$, then let v be the smallest alternative that is larger than x_0. Then $Y \cap G(z) = Y \cap G(x_0)$, so

$$\begin{aligned}
&\sum\{p_2(y)\colon y \in Y \cap G(z)\} \\
&= \sum\{p_2(y)\colon y \in Y \cap G(x_0)\} \\
&\geq \sum\{p_1(x)\colon x \in X \cap G(x_0)\} \\
&= \sum\{p_1'(x)\colon x \in X' \cap G(x_0)\} \\
&\geq \sum\{p_1'(x)\colon x \in X' \cap G(z)\}.
\end{aligned}$$

Moreover, because $|X' \cup Y| < |X \cup Y|$, our inductive hypothesis yields that $E_{u,p,i}(X') \leq E_{u,p,i}(Y)$. But X' was obtained from X by shifting the probability that was on x_0 up to the more preferred alternative w. Hence

$$\mathrm{E}_{u,p,i}(\mathrm{X}) < \mathrm{E}_{u,p,i}(\mathrm{X}');$$

so

$$\mathrm{E}_{u,p,i}(\mathrm{X}) < \mathrm{E}_{u,p,i}(\mathrm{Y}),$$

as desired. □

At this point, we can turn to a very pretty theorem of Allan Feldman (1980a). Feldman actually derived his theorem from an earlier result of Allan Gibbard.[18] The proof we give here is very different.

Theorem 4.3.2 (Feldman, 1980a). *In the context of linear ballots, if* n *is a positive integer and A is a set of three or more alternatives, then a non-imposed voting rule for* (*A*, n) *is non-manipulable in the sense of expected utility iff it is either a dictatorship or a duumvirate:*

Only the left-to-right direction requires proof, and this requires a few lemmas. In each, we assume that V is non-manipulable in the sense of expected utility, and non-imposed.

Lemma 4.3.3. *There exists a voter* i *such that, for every profile* $\boldsymbol{P}$ *we have* $top_{\mathrm{i}}(\boldsymbol{P}) \in V(\boldsymbol{P})$.

Proof: If V is non-manipulable in the sense of expected utility, then V cannot be manipulated by an optimist or a pessimist. Thus, the conclusion follows from the Duggan–Schwartz theorem. □

Lemma 4.3.4. *If a voter can unilaterally change the outcome of an election from* $V(\boldsymbol{P}) = Y$ *to* $V(\boldsymbol{P}') = X$, *then none of the following six conditions can hold (where, if it is voter* j *we are speaking of, we write* $max(X)$ *in place of* $max_{\mathrm{j}}(X, \boldsymbol{P})$*):*

(1) $max(X) > max(Y)$
(2) $max(X) = max(Y)$ *and* $|X| < |Y|$
(3) $max(X) = max(Y)$ *and* $|X| = |Y|$ *and* $max(X \Delta Y)$[19] $\in X$
(4) $min(X) > min(Y)$
(5) $min(X) = min(Y)$ *and* $|X| > |Y|$
(6) $min(X) = min(Y)$ *and* $|X| = |Y|$ *and* $min(X \Delta Y) \in Y$

[18] The result of Gibbard referred to here is not his original 1973 result, but a later version that is in quite a different context. We state this later result (from Gibbard, 1978) in Section 9.4.
[19] The symbol Δ denotes "symmetric difference." By definition, $\mathrm{A} \,\Delta\, \mathrm{B} = (\mathrm{A} - \mathrm{B}) \cup (\mathrm{B} - \mathrm{A})$.

Proof: In each of the six cases it suffices to find a $z \in X$ such that

$$|Y \cap G(z)|/|Y| < |X \cap G(z)|/|X|$$

We give an appropriate choice of z for each of the six cases, and leave the easy verification that each works to the reader.

(1) Let $z = \max(X)$
(2) Let $z = \max(X)$
(3) Let $z = \max(X \Delta Y)$
(4) Let $z = \min(X)$
(5) Let $z = \min(\{x \in A: x > \min(X)\})$
(6) Let $z = \min(\{x \in A: x > \min(X \Delta Y)\})$ □

Lemma 4.3.5. *If V is non-resolute, then there exist voters* i *and* j *such that, for every profile* $\boldsymbol{P}$, $\{top_i(\boldsymbol{P}), top_j(\boldsymbol{P})\} \subseteq V(\boldsymbol{P})$.

Proof: Using Lemma 4.3.4, we can choose i such that, for every profile **P**, $\text{top}_i(\mathbf{P}) \in V(\mathbf{P})$. Now Let k be the maximal size, over all **P**, of V(**P**) and note that $k \geq 2$ because V is not resolute. Define a new voting rule V′ as follows:

$$V'(\mathbf{P}) = \begin{cases} V(\mathbf{P}) & \text{if } |V(\mathbf{P})| < k \\ V(\mathbf{P}) - \{\text{top}_i(\mathbf{P})\} & \text{if } |V(\mathbf{P})| = k \end{cases}$$

V′ is clearly non-imposed and we claim that V′ is not manipulable by an optimist or a pessimist. To see this we'll argue by contradiction. So suppose that some voter (voter s) can unilaterally change an outcome from $V'(\mathbf{P}) = Y$ to $V'(\mathbf{P}') = X$ and that either $\max_s(X, \mathbf{P}) > \max_s(Y, \mathbf{P})$ or $\min_s(X, \mathbf{P}) > \min_s(Y, \mathbf{P})$. Let's call Y "old" if $V(\mathbf{P}) = Y$ also, and "new" if $V(\mathbf{P}) = Y \cup \{\text{top}_i(\mathbf{P})\}$, and the same for X. Notice that X and Y can't both be old, or else V would also be manipulable by an optimist or a pessimist, and thus manipulable in the sense of expected utility. Assume first that $\max_s(X, \mathbf{P}) > \max_s(Y, \mathbf{P})$.

Case 1: X is new and Y is old.

In this case, $\max_s(X \cup \{\text{top}_i(\mathbf{P})\}) > \max_s(Y, \mathbf{P})$, and so V is manipulable in the sense of expected utility by condition (1) in Lemma 4.3.4.

Case 2: X is old and Y is new.

Since $\text{top}_i(\mathbf{P}) \in X$, either $\max_s(X, \mathbf{P}) > \max_s(Y \cup \{\text{top}_i(\mathbf{P})\}, \mathbf{P})$, in which case V is manipulable in the sense of expected utility by condition (1) in Lemma 4.3.4, or else $\max_s(X, \mathbf{P}) = \max_s(Y \cup \{\text{top}_i(\mathbf{P})\}, \mathbf{P})$. In the latter case, we have $|Y \cup \{\text{top}_i(\mathbf{P})\}| = k > |X|$, and so V is manipulable in the sense of expected utility by condition (2) in 4.3.4.

Case 3: X is new and Y is new.
In this case we either have

$$\max_s(\mathrm{X} \cup \{\mathrm{top}_i(\mathbf{P})\}, \mathbf{P}) > \max_s(\mathrm{Y} \cup \{\mathrm{top}_i(\mathbf{P})\}, \mathbf{P})$$

or else

$$\max_s(\mathrm{X} \cup \{\mathrm{top}_i(\mathbf{P})\}, \mathbf{P}) = \mathrm{top}_i(\mathbf{P}) = \max_s(\mathrm{Y} \cup \{\mathrm{top}_i(\mathbf{P})\}, \mathbf{P}).$$

In the latter case, we also have $|\mathrm{X} \cup \{\mathrm{top}_i(\mathbf{P})\}| = k$ and $|\mathrm{Y} \cup \{\mathrm{top}_i(\mathbf{P})\}| = k$. But then we also have that

$$\max_s([\mathrm{X} \cup \{\mathrm{top}_i(\mathbf{P})\}]\Delta[\mathrm{Y} \cup \{\mathrm{top}_i(\mathbf{P})\}], \mathbf{P}) = \max_s(\mathrm{Y}, \mathbf{P}),$$

because $\max_s(\mathrm{X}, \mathbf{P}) > \max_s(\mathrm{Y}, \mathrm{P})$.

If $\min_s(\mathrm{X}, \mathbf{P}) > \min_s(\mathrm{Y}, \mathbf{P})$, the argument is completely analogous, using conditions 4, 5, and 6 of Lemma 4.3.4.

Because V′ is not manipulable by an optimist or a pessimist, we can apply the Duggan–Schwartz theorem to get a voter j such that $\mathrm{top}_j(\mathbf{P}) \in \mathrm{V}'(\mathbf{P})$ for every profile **P**. Note that $j \neq i$ as can be seen by a choice of **P** for which $|\mathrm{V}(\mathbf{P})| = k$, because this yields $\mathrm{top}_j(\mathbf{P}) \in \mathrm{V}'(\mathbf{P})$, but $\mathrm{top}_i(\mathbf{P}) \notin \mathrm{V}'(\mathbf{P})$. It now follows that for every profile **P**, $\{\mathrm{top}_i(\mathbf{P}), \mathrm{top}_j(\mathbf{P})\} \subseteq \mathrm{V}(\mathbf{P})$. □

Lemma 4.3.6. *If any voter changes his or her ballot by interchanging two adjacent alternatives, then the set of winners occurring on his or her ballot below both of these alternatives does not change.*

Proof: Assume that voter i's ballot $\mathrm{R}_i = \ldots ab \ldots y \ldots$ and that $y \in \mathrm{V}(\mathbf{P})$. Now suppose that $\mathbf{P}'$ is the result of voter i changing his or her ballot to $\mathrm{R}'_i = \ldots ba \ldots y \ldots$ and that $y \notin \mathrm{V}(\mathbf{P}')$. In R_i and R'_i the alternatives below a and b are the same and are ranked the same, so it makes sense to choose the lowest ranked $y \in \mathrm{V}(\mathbf{P}) \,\Delta\, \mathrm{V}(\mathbf{P}')$. Thus, for every $w < y$, $w \in \mathrm{V}(\mathbf{P})$ iff $w \in \mathrm{V}(\mathbf{P}')$.

Case 1: For every $w < y$, $w \notin \mathrm{V}(\mathbf{P})$ and $w \notin \mathrm{V}(\mathbf{P}')$.
In this case, $\min_i(\mathrm{V}(\mathbf{P}), \mathbf{P}) = y$ and $\min_i(\mathrm{V}(\mathbf{P}'), \mathbf{P}) > y$ in both ballots, and so this contradicts part 4 of Lemma 4.3.4.
Case 2: There exists some $w < y$ such that $w \in \mathrm{V}(\mathbf{P})$ and $w \in \mathrm{V}(\mathbf{P}')$.
In this case $\min_i(\mathrm{V}(\mathbf{P}), \mathbf{P}) = \min_i(\mathrm{V}(\mathbf{P}'), \mathbf{P})$. If $|\mathrm{V}(\mathbf{P})| \neq |\mathrm{V}(\mathbf{P}')|$, then we have a contradiction of part 5 of Lemma 4.3.4. If $|\mathrm{V}(\mathbf{P})| = |\mathrm{V}(\mathbf{P}')|$, then $\min_i(\mathrm{V}(\mathbf{P}) \,\Delta \mathrm{V}(\mathbf{P}'), \mathbf{P}) \in \mathrm{V}(\mathbf{P})$, and this contradicts part 6 of Lemma 4.3.4. □

Lemma 4.3.7. *V satisfies unanimity.*

Proof: Suppose not and choose $\mathbf{P}'$ such that $V(\mathbf{P}') = \{x\}$. One by one, have each voter change his or her ballot from what it was in $\mathbf{P}$ to what it was in $\mathbf{P}'$. At some point the winner becomes $\{x\}$, and the voter whose ballot change immediately preceded this has gained by obtaining his or her top choice as a singleton. □

Lemma 4.3.8. *V satisfies Pareto.*

Proof: Assume that every voter has x over y and that y is a winner. One by one, have each voter repeatedly interchange x with the alternative immediately above x on his or her ballot. By Lemma 4.3.6, y remains a winner. But at the end of this process, every voter has x on top of his or her ballot, and so by Lemma 4.3.7, the winner is $\{x\}$.

Lemma 4.3.9. *With* i *and* j *as in Lemma 4.3.5. $V(\boldsymbol{P}) = \{top_i(\boldsymbol{P}), top_j(\boldsymbol{P})\}$.*

Proof: For notational simplicity, assume $i = 1$ and $j = 2$, and let $x = \text{top}_1(\mathbf{P})$ and $y = \text{top}_2(\mathbf{P})$. Assume that $z \in V(\mathbf{P})$ where $z \neq x, y$. One by one we can move z to the top of ballots 3 through n and z remains a winner (or undoing such a change improves the max of the set of outcomes for that voter). Similarly we can move x up on these same ballots until it is just below z and then y up until it is just below x, all of this with z remaining a winner. Now have voter 2 move x up just below y on his or her ballot, and let $\mathbf{P}'$ denote the resulting profile. By Pareto, $V(\mathbf{P}') \subseteq \{x, y, z\}$ because every voter has x over every other alternative. Hence there are two cases to consider.

Case 1. $V(\mathbf{P}') = \{x, y\}$.

Voter 2 has $\max_2(V(\mathbf{P}), \mathbf{P}) = y$ and $\max_2(V(\mathbf{P}'), \mathbf{P}) = y$ and $|V(\mathbf{P})| > |V(\mathbf{P}')|$. It thus follows from part 2 of Lemma 4.3.4 that V is manipulable in the sense of expected utility.

Case 2. $V(\mathbf{P}') = \{x, y, z\}$.

In this case, we must have $V(\mathbf{P}) = \{x, y, z\}$ also, or we could argue as in Case 1. Suppose voter 2 exchanges x and y on his or her ballot. Pareto now rules out y as a winner. If the new winner is $\{x\}$, then voter 2 has improved his min from z to x. If the new winner is $\{x, z\}$, then we can take voter 2's true preferences to be the new ballot with x over y over z. Then his or her change back will change the set of winners from $\{x, z\}$ to $\{x, y, z\}$, so the min stays the same but the size of the set of winners increases. This contradicts part 5 of Lemma 4.3.4 and completes the proof of Lemma 4.3.8 and the proof of Feldman's theorem. □

4.4 Expected Utility Results

In this section, we consider expected utility results in the context where the probability function involved in breaking ties and comparing sets of alternatives is not specified by the procedure. Here, when comparing how a voter feels about two different sets X and Y, it is natural to assume that the probability function that he or she is using for X is quite independent of the probability function that he or she is using for Y. The point is that a voter might well feel that x has less of a chance of winning than y when both are members of one set X, and yet also feel that x has more of a chance of winning than y if both belong to some other set Y.[20]

Let us, at this point, establish some notation. For any set $X \subseteq A$, let P_X be the set of all probability functions mapping X to [0,1], and let P_X^+ be the set of all probability functions in P_X that are strictly positive. Let U_A^i be the set of all utility functions on A that represent voter i's preferences.

If $u \in U_A^i$ and $p \in P_X$, then the expected utility of X to voter i, denoted, $E_{u,p,i}(X)$, is given by:

$$E_{u,p,i}(X) = \sum \{p(x) \cdot u(x) : x \in X\}.$$

We would now like to say that voter i prefers set X to set Y iff $E_{u,p,i}(X) > E_{u,p,i}(Y)$. This, however, is an incomplete description unless we specify which quantifiers – universal or existential – are to be used for the probability functions and the utility functions, and, in some cases at least, the order of these quantifiers. For example, do we want to say that X is preferred to Y if the expected utility of X is greater than the expected utility of Y for *every* utility function and *every* pair of probability functions, or for *some* utility function and *some* pair of probability functions?

For the moment, we only consider probability functions that are strictly positive (i.e., in P_X^+). Moreover, for notational simplicity, we let u be a variable ranging over U_A and p be a variable ranging over $P_X^+ \times P_Y^+$. There are now eight possible definitions that naïvely suggest themselves:

(1) $X > Y$ iff $\forall u \, \forall p, E_{u,p,i}(X) > E_{u,p,i}(Y)$.
(2) $X > Y$ iff $\forall p \, \forall u, E_{u,p,i}(X) > E_{u,p,i}(Y)$.
(3) $X > Y$ iff $\exists u \, \forall p, E_{u,p,i}(X) > E_{u,p,i}(Y)$.
(4) $X > Y$ iff $\forall p \, \exists u, E_{u,p,i}(X) > E_{u,p,i}(Y)$.

[20] This, however, is not the only possibility. Ching and Zhou (2002) assume that a voter's probability functions on X and Y are derived from a probability function defined on the whole set A by using conditional probabilities. We do not consider this approach here. For related results, see Benoit (2002).

(5) X > Y iff $\exists p\, \forall u, E_{u,p,i}(X) > E_{u,p,i}(Y)$.
(6) X > Y iff $\forall p\, \exists u, E_{u,p,i}(X) > E_{u,p,i}(Y)$.
(7) X > Y iff $\exists u\, \exists p, E_{u,p,i}(X) > E_{u,p,i}(Y)$.
(8) X > Y iff $\exists p\, \exists u, E_{u,p,i}(X) > E_{u,p,i}(Y)$.

Clearly, we don't get eight distinct notions from these – for example, the first two are obviously equivalent, as are the last two. But there are less obvious equivalencies as well. Moreover, each of the eight definitions has a rather pleasing combinatorial equivalent in terms of the min and the max of the two sets being comparable. All of this is summarized in the following theorem.

Theorem 4.4.1 *Fix a linear ordering* P_i *(which we think of as voter* i*'s ballot) of the set A, and let X and Y be any two distinct non-empty subsets of A. Notationally, let* u *be a variable ranging over utility functions that represent* R_i*, and let* p *be a variable ranging over ordered pairs* (p_1, p_2) *where* p_1 *is a strictly positive probability function on X and* p_2 *is a strictly positive probability function on Y. Then:*

(1) The following are equivalent:
   (a) $min(X) \geq max(Y)$[21]
   (b) $\forall u\, \forall p, E_{u,p,i}(X) > E_{u,p,i}(Y)$
   (c) $\forall p\, \forall u, E_{u,p,i}(X) > E_{u,p,i}(Y)$
   (d) $\exists u\, \forall p, E_{u,p,i}(X) > E_{u,p,i}(Y)$
(2) The following are equivalent:
   (a) $max(X) \geq max(Y)$ *and* $min(X) \geq min(Y)$
   (b) $\exists p\, \forall u, E_{u,p,i}(X) > E_{u,p,i}(Y)$
(3) The following are equivalent:
   (a) $max(X) > max(Y)$ *or* $min(X) > min(Y)$
   (b) $\forall p\, \exists u, E_{u,p,i}(X) > E_{u,p,i}(Y)$
(4) The following are equivalent:
   (a) $max(X) > min(Y)$
   (b) $\forall u\, \exists p, E_{u,p,i}(X) > E_{u,p,i}(Y)$
   (c) $\exists u\, \exists p, E_{u,p,i}(X) > E_{u,p,i}(Y)$
   (d) $\exists p\, \exists u, E_{u,p,i}(X) > E_{u,p,i}(Y)$

Proof: Assume that (1a) holds and the utility function u and the probability functions on X and Y are arbitrary. Then

$$E_{u,p,i}(X) \geq u(\min(X)) \geq u(\max(Y)) \geq E_{u,p,i}(Y).$$

[21] At first blush, it looks as if there might be a problem with the use of "≤" as opposed to "<" because min(X) = max(Y) is satisfied if both X and Y are the same singleton set. However, the assumption in Theorem 4.4.1 is that X and Y are *distinct* sets.

Moreover, if $X - Y \neq \emptyset$, then the first inequality is strict, and if $Y - X \neq \emptyset$, then the last inequality is strict.

Clearly (1b) implies (1c) implies (1d). To see that (1d) implies (1a), assume that $\min(X) < \max(Y)$ and assume that u is an arbitrary utility function that represents P_i. We will produce strictly positive probability functions $p_1\colon X \to [0,1]$ and $p_2\colon Y \to [0,1]$ such that, letting $p = (p_1, p_2)$, we have $E_{u,p,i}(X) < E_{u,p,i}(Y)$. (We really only need $\leq$ here, but the stronger conclusion is required for part (4) of the theorem.)

If not for the fact that the probability functions have to be strictly positive, we could set $p_1(\min(X)) = 1$ and $p_2(\max(Y)) = 1$, and, with $p = (p_1, p_2)$, have $E_{u,p,i}(X) = u(\min(X)) < u(\max(Y)) = E_{u,p,i}(Y)$. However, we can still make this argument work by choosing $p_1(\min(X))$ and $p_2(\max(Y))$ very close to 1. So choose $q \in (0, 1)$ close enough to 1 so that $(1 - q) \cdot u(\max(X)) < q \cdot [u(\max(Y)) - u(\min(X))]$. Now let p_1 and p_2 be any strictly positive probability functions on X and Y respectively such that $p_1(\min(X)) = q$ and $p_2(\max(Y)) = q$. Then

$$\begin{aligned}
E_{u,p,i}(X) &\leq p_1(\min(X)) \cdot u(\min(X)) + (1 - p_1(\min(X)) \cdot u(\max(X)) \\
&= q \cdot u(\min(X)) + (1 - q) \cdot u(\max(X)) \\
&< q \cdot u(\min(X)) + q \cdot [u(\max(Y)) - u(\min(X))] \\
&= q \cdot u(\max(Y)) \\
&= p_2(\max(Y)) \cdot u(\max(Y)) \\
&\leq E_{u,p,i}(Y).
\end{aligned}$$

This completes that proof that (1a)–(1d) are equivalent.

For (2), assume first that 2(a) holds, so we have $\max(X) \geq \max(Y)$ and $\min(X) \geq \min(Y)$. Set $p_1(\max(X)) = 3/4$ and $p_2(\min(Y)) = 3/4$, with p_1 and p_2 assigned arbitrarily (but positive) on the rest of X and Y respectively. Let $p = (p_1, p_2)$. Assume that u is arbitrary. Then

$$\begin{aligned}
E_{u,p,i}(X) &\geq p_1(\max(X)) \cdot u(\max(X)) + (1 - p_1(\max(X))) \cdot u(\min(X)) \\
&= (3/4) \cdot u(\max(X)) + (1/4) \cdot u(\min(X)) \\
&\geq (3/4) \cdot u(\max(Y)) + (1/4) \cdot u(\min(Y)) \\
&\geq (3/4) \cdot u(\min(Y)) + (1/4) \cdot u(\max(Y)) \\
&= p_2(\min(Y)) \cdot u(\min(Y)) + (1 - p_2(\min(Y))) \cdot u(\max(X)) \\
&\geq E_{u,p,i}(Y).
\end{aligned}$$

If X has three or more elements, the first inequality is strict, and if Y has three or more elements, the last is. Otherwise, we can't have max(X) = max(Y) and min(X) = min(Y), so the middle inequality is strict.

Now assume that (2a) fails; we need to show that $\forall p\ \exists u\ \mathrm{E}_{u,p,i}(\mathrm{X}) \leq \mathrm{E}_{u,p,i}(\mathrm{Y})$. Assume that the probability functions on X and Y are arbitrary. We will construct u such that $\mathrm{E}_{u,p,i}(\mathrm{X}) < \mathrm{E}_{u,p,i}(\mathrm{Y})$. Again, we really only need $\leq$ for the proof here, but we'll need the stronger result for the proof in (3) below.

Case 1: max(X) < max(Y)
Assume that u is assigned arbitrarily for all $z < \max(\mathrm{Y})$, and let $p = \mathrm{prob}(\max(\mathrm{Y}))$. Set $u(\max(\mathrm{Y}))$ large enough so that

$$u(\max(\mathrm{Y})) > [u(\max(\mathrm{X})) - (1-p) \cdot u(\min(\mathrm{Y}))]/p.$$

Then

$$\begin{aligned} \mathrm{E}_{u,p,i}(\mathrm{X}) &\leq u(\max(\mathrm{X})) \\ &< p \cdot u(\max(\mathrm{Y})) + (1-\mathrm{p}) \cdot u(\min(\mathrm{Y})) \\ &\leq \mathrm{E}_{u,p,i}(\mathrm{Y}). \end{aligned}$$

Case 2: min(X) < min(Y)
Assume that u is assigned arbitrarily for all $z > \min(\mathrm{X})$, and let $p = \mathrm{prob}(\min(\mathrm{X}))$. Set $u(\min(\mathrm{X}))$ small enough so that

$$u(\min(\mathrm{X})) < [u(\min(\mathrm{Y})) - (1-p) \cdot u(\max(\mathrm{X}))]/p.$$

Then

$$\begin{aligned} \mathrm{E}_{u,p,i}(\mathrm{X}) &\leq p \cdot u(\min(\mathrm{X})) + (1-p) \cdot u(\max(\mathrm{X})) \\ &< u(\min(\mathrm{Y})) \\ &\leq \mathrm{E}_{u,p,i}(\mathrm{Y}). \end{aligned}$$

This completes that proof that (2a) and (2b) are equivalent.

For (3), assume first that (3a) holds. Then max(X) > max(Y) or min(X) > min(Y). But our proof of the equivalencies in (2) showed that this implies that $\forall p\ \exists u,\ \mathrm{E}_{u,p,i}(\mathrm{X}) > \mathrm{E}_{u,p,i}(\mathrm{Y})$ and so (3b) holds. Conversely, if (3a) fails, then max(X) ≤ max(Y) and min(X) ≤ min(Y), and our proof of the equivalencies in (2) showed that this implies that $\exists p\ \forall u,\ \mathrm{E}_{u,p,i}(\mathrm{X}) > \mathrm{E}_{u,p,i}(\mathrm{Y})$, and so (3b) fails also.

For (4), assume first that (4a) holds. Then by our proof of the equivalencies in (1), $\forall u\ \exists p,\ \mathrm{E}_{u,p,i}(\mathrm{X}) > \mathrm{E}_{u,p,i}(\mathrm{Y})$, and so (4b) holds. It is now trivial to see that (4b) implies (4c) implies (4d). Finally, if (4d) holds then $\exists p\ \exists u,\ \mathrm{E}_{u,p,i}(\mathrm{X}) > \mathrm{E}_{u,p,i}(\mathrm{Y})$, and so it is not the case that $\forall p\ \forall u,\ \mathrm{E}_{u,p,i}(\mathrm{X}) \leq \mathrm{E}_{u,p,i}(\mathrm{Y})$. Hence, it is not the case that $\forall p\ \forall u$, $E_{u,p,i}(X) < E_{u,p,i}(Y)$, so (1c) fails. Thus, (1a) fails, so min(Y) < max(X). This completes the proof of Theorem 4.4.1. □

In terms of strength, the four notions in Theorem 4.4.1 form a diamond:

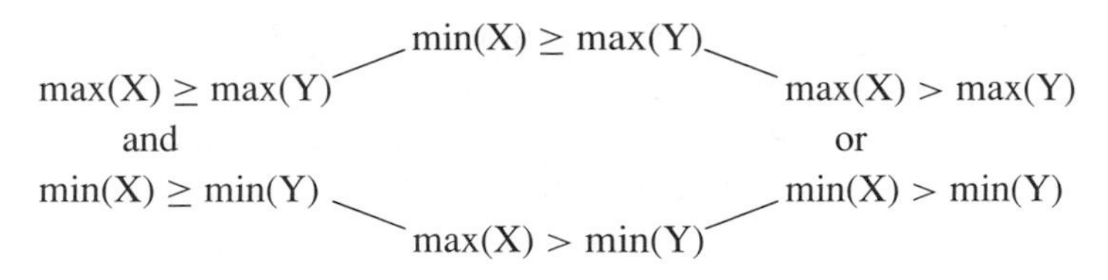

Each of these four notions provides a way to order sets of alternatives based on the kind of preference ordering that we use as ballots. Thus, we have four chances to ask if there exists a "manipulability theorem" asserting that every reasonable voting rule can be manipulated in the sense of some voter's ability to disingenuously change the election outcome from one set X to another set Y that he or she prefers to X.

It turns out that two of the notions yield theorems (in fact, the Gibbard–Satterthwaite theorem and the Duggan–Schwartz theorem), while the other two notions don't. The embellished picture is as follows:

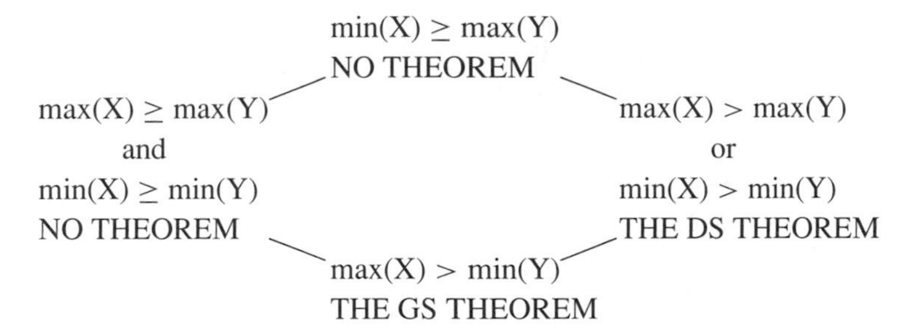

The verification of these assertions is quite easy, and goes as follows. For the two "no theorems" results, we let V be the Condorcet voting rule. We claim that a voter can't unilaterally change the outcome of an election without making the min of the set of winners strictly worse, or the max of the set of winners strictly worse.

To see this, notice that if a voter changes his or her ballot so that the winner changes from $\{a\}$ to A, then a was not on the bottom of his or her ballot, or else it would remain a Condorcet winner no matter what ballot change the voter made. Hence, the min gets worse. Similarly, if the voter changes his or her ballot so that the winner changes from A to $\{a\}$, then a was not on top of his or her ballot and so the max gets worse. Finally, if a voter changes his or her ballot so that the winner changes from $\{a\}$ to $\{b\}$, then he or she had a over b to begin with and reversed this to effect the change. Hence the min and max both get worse.

The fact that the Duggan–Schwartz theorem corresponds to the notion in the box on the right is obvious, and the fact the box on the bottom corresponds to the Gibbard–Satterthwaite theorem follows from Corollary 3.1.13.

It turns out that if we allow the probability functions to take on the value zero, then the analogue of Theorem 4.4.1 is somewhat less interesting (although easier to prove). However, it does add a fifth possibility, and so we state it here and leave the proof as an exercise for the reader.

Proposition 4.4.2. *Fix a linear ordering P_i of the set A, and let X and Y be any two distinct subsets of A. Notationally, let* u *be a variable ranging over utility functions that represents R_i, and let* p *be a variable ranging over ordered pairs* (p_1, p_2) *where* p_1 *is a probability function on X and* p_2 *is a probability function on Y, where we allow probabilities in both cases to be zero. Then,*

(1) The following are equivalent:
- *(a)* $min(X) > max(Y)$
- *(b)* $\forall u\ \forall p,\ E_{u,p,i}(X) > E_{u,p,i}(Y)$
- *(c)* $\forall p\ \forall u,\ E_{u,p,i}(X) > E_{u,p,i}(Y)$
- *(d)* $\exists u\ \forall p,\ E_{u,p,i}(X) > E_{u,p,i}(Y)$
- *(e)* $\forall p\ \exists u,\ E_{u,p,i}(X) > E_{u,p,i}(Y)$

(2) The following are equivalent:
- *(a)* $max(X) > min(Y)$
- *(b)* $\forall u\ \exists p,\ E_{u,p,i}(X) > E_{u,p,i}(Y)$
- *(c)* $\exists u\ \exists p,\ E_{u,p,i}(X) > E_{u,p,i}(Y)$
- *(d)* $\exists p\ \exists u,\ E_{u,p,i}(X) > E_{u,p,i}(Y)$
- *(e)* $\exists p\ \forall u,\ E_{u,p,i}(X) > E_{u,p,i}(Y)$

There is one final comment that we should make. The astute reader might have noticed that at least one of our ways to order sets of alternatives is somewhat counterintuitive. That is, if we declare X > Y to mean that $\exists p\ \forall u$, $E_{u,p,i}(X) > E_{u,p,i}(Y)$, then this says we have X > Y whenever $\max(X) \geq \max(Y)$ and $\min(X) \geq \min(Y)$. But this says that if I rank a over b over c over d, then I might prefer the set $\{a, c, d\}$ to the set $\{a, b, d\}$. At first blush, this seems impossible, but suppose that our voting rule specifies that ties must be broken by first using the sequential pairwise rule, with alternatives in the order d, c, b, a, and then, if ties still remain, some other device is used (which device is actually used here is irrelevant for the present argument). Suppose that there are six other voters and their ballots, together with mine (the one on the far left), are as follows:

a	a	a	b	b	d	d
b	c	c	c	c	b	b
c	d	d	d	d	a	a
d	b	b	a	a	c	c

If the set of winners is $\{a, b, d\}$, then sequential pairwise voting will first pit d against b (with d winning 4 to 3), then d against a with d again winning 4 to 3). But if the set of winners is $\{a, c, d\}$, then sequential pairwise voting will first pit d against c (with c winning 5 to 2) and then c against a (with a winning 5 to 2). Thus, I am much better off with an initial tie among my first, third, and fourth choices – obtaining my top choice a from $\{a, c, d\}$ – than with an initial tie among my first, second, and third choices – obtaining my bottom choice d from $\{a, b, d\}$.

5
Social Choice Functions

5.1 The Barberá–Kelly Theorem

Throughout this chapter, V denotes a social choice function. We assume there are three or more alternatives in the set A, and that V is defined for every two-element agenda. Except where mentioned otherwise, we restrict attention to questions of manipulability in the context of linear ballots.

Manipulability of social choice functions was touched on in Section 2.4. The difference between what was done there and what is done here is twofold: Here, we only consider ballot manipulability – not agenda manipulability – and we do not restrict ourselves to resolute procedures.

But this means that we must now confront the same issue that arose in earlier chapters: What does it mean to say "there is a situation in which voter i can secure a better outcome by submitting an insincere ballot?" Aside from revisiting the problem of deriving preferences for sets of alternatives from a voter's ranking of individual alternatives, there is a new aspect of "situation" to be dealt with here, and we begin by dispensing with it.

Suppose, then, that we have a social choice function V that is (ballot) manipulable in some sense. Does this mean that V is manipulable for *every* agenda v or does it mean that V is manipulable for *some* agenda v? It is, in fact, the latter notion with which we work, because saying that V is manipulable for every agenda v would mean, in particular, that V is manipulable when $v = \mathrm{A}$, and this puts us back in the context of voting rules – a topic we've finished with for the moment.

The notion of manipulability that we use in this chapter is weak-dominance manipulability, and, in point of fact, the primary (but not only) instance we need at the moment arises from the observation that if voter i prefers x to y, then he or she will obviously prefer $\{x\}$ to $\{x, y\}$ and $\{x, y\}$ to $\{y\}$.

Definition 5.1.1. In the context of linear or non-linear ballots, a social choice function V is *weak-dominance manipulable on the agenda v* if there exists a profile $\mathbf{P} = (R_1, \ldots, R_n)$, which we think of as giving the true preferences of the n voters, another ballot Q_i, which we think of as a disingenuous ballot from voter i such that, letting $\mathbf{P}' = (R_1, \ldots, R_{i-1}, Q_i, R_{i+1}, \ldots, R_n)$,

$$\forall x \in V(\mathbf{P}')(v)\ \forall y \in V(\mathbf{P})(v),\ xR_i y \quad \text{and} \quad \exists x \in V(\mathbf{P}')(v)\ \exists y \in V(\mathbf{P})(v),\ xP_i y$$

V is *weak-dominance manipulable* (or *manipulable in the sense of weak dominance*) if it is weak-dominance manipulable on some agenda.

The conclusion in Definition 5.1.1 is, for two-element agendas, equivalent to saying that V cannot be manipulated by an optimist or a pessimist, as in the last chapter. We also need an analog of two other definitions from that chapter.

Definition 5.1.2. In the context of linear or non-linear ballots, a social choice function V is *pairwise non-imposed* if for every alternative x and every alternative y, there exists at least one profile **P** such that $V(\mathbf{P})(\{x, y\}) = \{x\}$.

Definition 5.1.3. If V is a social choice function, then voter i is a *pairwise nominator* (*for V*) if, for every ordered pair (x, y) of distinct alternatives, and for every profile **P** with $xP_i y$, we have $x \in V(\mathbf{P})(\{x, y\})$.

Intuitively, voter i is "nominating" x in the sense of being able to unilaterally ensure that x is among the winners, a power analogous to what every voter has with the omninominator voting rule. At the same time, we can interpret this power as the ability to "block y" or "veto y" in the sense of preventing y from being the *unique* choice from the two-element agenda $\{x, y\}$.

At this point, we want to step back and ask what, if anything, distinguishes a social choice function from a collection of completely unrelated voting rules, one chosen for each agenda v. One could, for example, use majority rule on one two-element agenda $\{x, y\}$ and a dictatorship on another two-element agenda $\{z, w\}$. Or one could use the Borda count on agendas of size three and the Hare system on agendas of size four.

In point of fact, a social choice function V acquires coherence only in the presence of one of two kinds of consistency requirements:

(1) Consistency across agendas of a fixed size; for example, asserting that, for every profile **P**, if $V(\mathbf{P})(\{x, y\}) = \{x\}$ and $V(\mathbf{P})(\{y, z\}) = \{y\}$, then $V(\mathbf{P})(\{x, z\}) = \{z\}$.

(2) Consistency from agendas of one size to those of another size, asserting, for example, that for every profile **P**, $x \in \mathrm{V}(\mathbf{P})(\{x, y, z\})$ iff $x \in \mathrm{V}(\mathbf{P})(\{x, y\}) \cap \mathrm{V}(\mathbf{P})(\{x, z\})$.

The main result in this section (the Barberá–Kelly theorem) is based on a consistency requirement of the first kind called "quasitransitivity." The MacIntyre–Pattanaik theorem in Section 5.4 is based on a consistency requirement of the second kind that we introduce later. Both of these requirements are defined in terms of the following notion.

Definition 5.1.4. If V is a social choice function and **P** is a profile, the *base relation for **P*** is the binary relation $\mathrm{R}_{\mathbf{P}}$ defined on A by

$$x\mathrm{R}_{\mathbf{P}}y \text{ iff } x \in \mathrm{V}(\mathbf{P})(\{x, y\}).$$

The derived strict relation $\mathrm{P}_{\mathbf{P}}$ is given by:

$$x\mathrm{R}_{\mathbf{P}}y \text{ iff } \mathrm{V}(\mathbf{P})(\{x, y\}) = \{x\}.$$

Notice that the base relation is always reflexive because $\mathrm{V}(\mathbf{P})(\{x\}) \neq \emptyset$ and complete because $\mathrm{V}(\mathbf{P})(\{x, y\}) \neq \emptyset$. With the definition of the base relation, we can now introduce the consistency-across-two-element-agendas requirement known as quasitransitivity.

Definition 5.1.5. A social choice function is *quasitransitive* if, for every profile **P**, the strict relation $\mathrm{P}_{\mathbf{P}}$ derived from the base relation $\mathrm{R}_{\mathbf{P}}$ is transitive.

The adjective "transitive" is reserved in this context for the relation $\mathrm{R}_{\mathbf{P}}$ itself. It is easy to see that the transitivity of $\mathrm{R}_{\mathbf{P}}$ implies the transitivity of $\mathrm{P}_{\mathbf{P}}$, but the converse fails. For example, if V is a social choice function that agrees with the unanimity rule on every two-element agenda, then the relation $\mathrm{R}_{\mathbf{P}}$ can fail to be transitive for some profiles **P** (Exercise), but V is quasitransitive as well as pairwise-non-imposed and non-manipulable in the sense of weak dominance on two-element agendas (Exercise). Notice also that every voter is a pairwise nominator for V in this case, something that foreshadows the upcoming theorem.

We attribute the main result of this section to Barberá and Kelly, as it is heavily based on two very similar results – one proved by Salvador Barberá in his 1975 Ph.D. thesis and published two years later in the *Journal of Economic Theory* (Barberá, 1977a), and the other proved by Jerry Kelly and published in *Econometrica* (Kelly, 1977) the same year that Barberá's article appeared.

The proof we give of this result is also quite different from that of either Barberá or Kelly and follows the outline of the proofs we gave for both the

Gibbard–Satterthwaite theorem in Chapter 3 and the Duggan–Schwartz theorem in Chapter 4. For what follows, we only need the social choice function V to be defined for two-element agendas.

Theorem 5.1.6 (The Barberá–Kelly Theorem for Linear Ballots). *In the context of linear ballots, if* n *is a positive integer and A is a set of three or more alternatives, then any social choice function for* (A, n) *that is quasitransitive, pairwise non-imposed, and non-manipulable in the sense of weak dominance on two-element agendas has a pairwise nominator.*

Our starting point in the proof is to introduce two definitions that mimic what we did in Chapter 3 with the Gibbard–Satterthwaite theorem and in Chapter 4 with the Duggan–Schwartz theorem.

Definition 5.1.7. If X is a set of voters, and a and b are two distinct alternatives in the set A, then *X can use* a *to block* b, denoted $a\text{X}b$, if, for every profile **P** in which all the voters in X rank a over b on their ballots, $\text{V}(\mathbf{P})(\{a, b\}) \neq \{b\}$. The set X is a *nominating set* if $a\text{X}b$ for every distinct pair a, b of alternatives in A.

Definition 5.1.8. V satisfies *pairwise down-monotonicity for singleton winners* provided that, for every profile **P** and every two-element agenda v, if $|\text{V}(\mathbf{P})(v)| = 1$, and $\mathbf{P}'$ is the profile obtained from **P** by having one voter move one losing alternative down one spot on his ballot, then $\text{V}(\mathbf{P}')(v) = \text{V}(\mathbf{P})(v)$.

Notice that if V satisfies pairwise down-monotonicity for singleton winners, v is a two-element agenda, and $\text{V}(\mathbf{P})(v) = \{x\}$, then $\text{V}(\mathbf{P}')(v) = \{x\}$ whenever $\mathbf{P}'$ is derived from **P** by having several voters move several alternatives down several slots on their ballots.

Lemma 5.1.9. *Every social choice function that is not manipulable in the sense of weak-dominance on two-element agendas satisfies pairwise down-monotonicity for singleton winners.*

Proof: Assume that pairwise down-monotonicity for singleton winners fails. Then there exist two profiles **P** and $\mathbf{P}'$, a two-element agenda $v = \{a, b\}$, and two alternatives x and y such that:

(1) In **P**, voter i ranks y directly over x, $\text{V}(\mathbf{P})(v) = \{a\}$, and $a \neq y$ (that is, y is the losing alternative that voter i will be moving down).
(2) $\mathbf{P}'$ differs from **P** only in that voter i has interchanged the position of x and y on his or her ballot, and yet $b \in \text{V}(\mathbf{P}')(v)$.

Assuming (1) and (2), we will show that the system can, in fact, be manipulated on the two-element agenda $\{a, b\}$.

Case 1: b is over a on voter i's ballot in **P**.

In this case, we can regard voter i's ballot in **P** as giving his or her true preferences, and see that voter i's disingenuous submission of Q_i improves the outcome (according to his true preferences) from $\{a\}$ to $\{b\}$ or $\{a\}$ to $\{a, b\}$.

Case 2: a is over b on voter i's ballot in $\mathbf{P}'$.

In this case, we can regard voter i's ballot in $\mathbf{P}'$ as giving his or her true preferences, and see that voter i's disingenuous submission of P_i improves the outcome (according to his or her true preferences) from $\{b\}$ to $\{a\}$ or $\{a, b\}$ to $\{a\}$.

Case 3: Otherwise.

In this case, a is over b on voter i's ballot in **P**, and b is over a on voter i's ballot in $\mathbf{P}'$. But this means that not only do we have $\{y, x\} = \{a, b\}$, but $y = a$ and $x = b$. Thus, our assumption that $a \neq y$ implies that Case 3 cannot occur. This proves the lemma. □

At this point, we are ready to set up for another appeal to Proposition 3.4.2. For the remainder of the proof, we assume that V is a fixed social choice function that is quasitransitive, pairwise non-imposed, and non-manipulable on two-element agendas, and therefore satisfies pairwise down-monotonicity for singleton winners.

Lemma 5.1.10 (The Existence Lemma). *Suppose that X is a set of voters and that* a *and* b *are two distinct alternatives in A. Then, in order to show that* a*X*b, *it suffices to produce one profile **P** for which:*

(1) {a, b} *is a top set,*
(2) everyone in X ranks a *over* b,
(3) everyone else ranks b *over* a, *and*
(4) a ∈ *V*(***P***)({a, b}).

Proof: Suppose we have such a profile **P** but aXb fails. Then we also have a profile $\mathbf{P}'$ in which everyone in X ranks a over b and $\mathrm{V}(\mathbf{P}')(\{a, b\}) = \{b\}$. Using pairwise down-monotonicity for singleton winners, we can convert $\mathbf{P}'$ into the profile **P** assumed to exist, and have $\mathrm{V}(\mathbf{P})(\{a, b\}) = \{b\}$. But this is a contradiction because $a \in \mathrm{V}(\mathbf{P})(\{a, b\})$. □

Lemma 5.1.11 (The Splitting Lemma). *Suppose X is a set of voters, and that* a, b, *and* c *are distinct alternatives in A. Assume also that* a*X*b *and that X is* *partitioned into disjoint sets Y and Z (one of which may be empty). Then either* a*Y* c *or* cZb.

Proof: Consider the profile **P** in which every voter in Y has a first, b second, and c third; every voter in Z has c first, a second, and b third; and everyone else (i.e., those in N–X) has b first, c second, and a third. We can picture these ballots as follows:

Y	Z	N–X
a	c	b
b	a	c
c	b	a
.	.	.
.	.	.
.	.	.

Because $a\text{X}b$ and everyone in X = Y ∪ Z has a over b, $\text{V}(\mathbf{P})(\{a, b\}) \neq \{b\}$. Now if $\text{V}(\mathbf{P})(\{b, c\}) = \{b\}$ and $\text{V}(\mathbf{P})(\{c, a\}) = \{c\}$, then $\text{V}(\mathbf{P})(\{a, b\}) = \{b\}$ by quasitransitivity. Hence, we have two cases:

Case 1: $\text{V}(\mathbf{P})(\{b, c\}) \neq \{b\}$.
Then $\text{V}(\mathbf{P})(\{b, c\}) = \{c\}$ or $\{c, b\}$. For each voter in Y, we one by one move a just below c. As we do this, c remains a winner or else such a change would represent an instance of weak-dominance manipulability. Similarly, we can one by one move a below b for each voter in Z, and have c remain a winner lest undoing such a change would be an instance of weak-dominance manipulability. Thus, $c\text{Z}b$ by Lemma 5.1.10.
Case 2: $\text{V}(\mathbf{P})(\{c, a\}) \neq \{c\}$.
As in case 1, we can make $\{a, c\}$ a top set so that $\text{V}(\mathbf{P})(\{a, c\}) = \{a\}$ or $\{a, c\}$, so $a\text{Y}c$ by Lemma 5.1.10. again. □

Lemma 5.1.12. *If X is a nominating set, then there exists a voter* i ∈ *X such that* {i} *is a nominating set. Moreover, N is a nominating set, so there exists a voter* i *such* i *is a nominator for V.*

Proof: The result follows immediately from Proposition 3.4.2. and Lemma 5.1.11, if we can establish that for every pair a and b of alternatives, $a\text{N}b$ holds and $a\emptyset b$ fails.

Suppose, for contradiction, that **P** is a profile in which every voter has a over b on his or her ballot, but $\text{V}(\mathbf{P})(\{a, b\}) = \{b\}$. Choose a profile $\mathbf{P}'$ such that $\text{V}(\mathbf{P}')(\{a, b\}) = \{a\}$. Now, using pairwise down-monotonicity for singleton winners, we can first move b to the bottom of every ballot in $\mathbf{P}'$ and then repeat this for each of the other losing alternatives in some fixed order. Similarly, we can move all alternatives other than a and b to the bottom, in this same fixed order, of all the ballots in $\mathbf{P}$. But then we have identical profiles with two different election outcomes.

To see that $a\emptyset b$ fails, choose a profile $\mathbf{P}$ such that $V(\mathbf{P})(\{a, b\}) = \{b\}$. Then every voter in $\emptyset$ (there are none) has a over b, but $V(\mathbf{P})(\{a, b\}) = \{b\}$. This completes the proof of Lemma 5.1.12. and Theorem 5.1.6. □

There is an extension of Theorem 5.1.6 due to Barberá that was originally based on a version of Arrow's impossibility theorem obtained (but never published) by Allan Gibbard. The starting point is our observation following Definition 5.1.5 that the unanimity rule U for pairs satisfies quasitransitivity, pairwise non-imposition, and non-manipulability in the sense of weak dominance for two-element agendas. The procedure U, however, is not the only voting procedure for pairs that satisfies these three properties. One could, for example add additional voters and make them dummies in the technical sense: their ballots are completely ignored. The resulting procedure is an oligarchy, and we need the pairwise version of this.

Definition 5.1.13. If O is a set of voters, then the corresponding *oligarchic procedure for pairs* is the social choice function V_O defined for every two-element agenda as follows:

$$V_O(\{x, y\}) = \begin{cases} \{x\} & \text{if } \forall i \in O \ xP_i y \\ \{y\} & \text{if } \forall i \in O \ yP_i x \\ \{x, y\} & \text{otherwise} \end{cases}$$

That is, $x \in V_O(\{x, y\})$ iff there exists a voter $i \in O$ such that $xP_i y$.

A social choice function is said to be a *pairwise oligarchy* if there exists a set O of voters such that $V|[A]^2 = V_O|[A]^2$.

Notice that if V is a pairwise oligarchy, then each voter in O (each "oligarch") is a pairwise nominator in the sense of Definition 5.1.3. Thus, each voter in O is guaranteed that his or her top choice (between the two alternatives x and y) is always among the winners. This, too, is reminiscent of the Duggan–Schwartz theorem. Moreover, the set O is a "pairwise dictating set" in the sense that if all the voters in O place x over y, then $V(\mathbf{P})(\{x, y\}) = \{x\}$.

With these preliminaries at hand, we can now show that the oligarchic procedures for pairs (with arbitrary extensions to agendas of size three or more) are precisely the ones that satisfy quasitransitivity, pairwise non-imposition, and non-manipulability in the sense of weak dominance for two-element agendas – and thus that the unanimity procedure U for pairs is completely characterized by these three properties plus the absence of dummies.

Corollary 5.1.14 (Barberá, 1977a). *In the context of linear ballots, if* n *is a positive integer and A is a set of three or more alternatives, then a social choice function for* (A, n) *is quasitransitive, pairwise non-imposed, and non-manipulable in the sense of weak dominance on two-element agendas iff it is a pairwise oligarchy.*

Proof: The proof that a pairwise oligarchy is quasitransitive, pairwise non-imposed, and non-manipulable for two-element agendas was outlined in the last section. For the converse, we know from Theorem 5.1.6 that if V satisfies these three conditions, then there is at least one nominator. We claim that if we let O be the set of nominators, then O is an oligarchy. To verify this, it suffices to show that if **P** is any profile in which every voter in O places x over y then V(**P**)($\{x, y\}$) = $\{x\}$.

Suppose not. Let Y be the set of all voters who have y over x on their ballot. Notice that O ∩ Y = ∅, that Y ≠ ∅, and yYx by the existence lemma (where non-manipulability can be used to make $\{x, y\}$ a top set). But Lemma 5.1.12 now shows that there exists a voter $i \in$ Y such that i is a nominator, and this is a contradiction. □

A voter is a *pairwise dummy* for a social choice function V if his or her vote can never affect V(**P**)(v) for any profile **P** and any two-element agenda v. The fact that a pairwise oligarchy with no dummies must agree with the unanimity rule on every two-element agenda immediately yields the following.

Corollary 5.1.15. *In the context of linear ballots, if* n *is a positive integer and A is a set of three or more alternatives, then a social choice function for* (A, n) *is quasitransitive, pairwise non-imposed, non-manipulable in the sense of weak dominance on two-element agendas, and has no pairwise dummies iff it agrees with the unanimity rule on every two-element agenda.*

5.2 Ties in the Ballots

Building on Barberá's work (1977a, p. 274), we show that the conclusion of Corollary 5.1.14 remains intact even if we allow ties in the ballots.

Theorem 5.2.1. *In the context of non-linear ballots, if* n *is a positive integer and A is a set of three or more alternatives, then a social choice function for*

(A, n) *that is quasitransitive, pairwise non-imposed, and non-manipulable in the sense of weak-domination for two-element agendas is a pairwise oligarchy.*

Proof: If we consider V restricted to linear ballots, then this restriction is still quasitransitive, pairwise non-imposed and non-manipulable for two-element agendas. Hence, by Corollary 5.1.14, V restricted to linear ballots is an oligarchy. Hence, the desired result will follow from the following two claims.

Claim 1. If voter i is a nominator for linear ballots, then he or she is still a nominator if we allow ties in the ballots.

Proof: Suppose **P** is a profile (having ties in some of the ballots) for which $x\mathrm{P}_i y$ holds, but $\mathrm{V}(\mathbf{P})(\{x, y\}) = \{y\}$. Let T be any strict linear ordering of the set A of alternatives such that $y > x$ in T. Let $\mathbf{P}'$ be the result of ballot-by-ballot breaking all the ties according to T. Because we still have $x\mathrm{P}_i y$ in $\mathbf{P}'$, we know that $x \in \mathrm{V}(\mathbf{P}')(\{x, y\})$. Choose j such that breaking the ties in voter j's ballot first resulted in x becoming a winner.

Case 1: $y\mathrm{R}_j x$.

Then either $y\mathrm{P}_j x$ or $y\mathrm{I}_j x$ – either way we have $y\mathrm{P}_j x$ because $y > x$ in T. But then, if we regard R_j' to be voter j's true preferences, this is an instance of manipulation of a two-element agenda.

Case 2: $x\mathrm{P}_j y$.

In this case, we get an instance of manipulation of a two-element agenda if we regard R_j as voter j's true preferences. This proves Claim 1.

Claim 2. If O is a nominating set for linear ballots, then O is a nominating set with ties allowed in the ballots.

Proof: Suppose **P** is a profile in which everyone in O ranks x over y, but $y \in \mathrm{V}(\mathbf{P})(\{x, y\})$. Again, let T be any strict linear ordering of the set A of alternatives such that $y > x$ in T, and let $\mathbf{P}'$ be the result of ballot-by-ballot breaking all the ties according to T. Because we still have that everyone in O ranks x over y, we know that $\mathrm{V}(\mathbf{P}')(\{x, y\}) = \{x\}$. Choose j such that breaking the ties in voter j's ballot first resulted in $\{x\}$ becoming the winner. As before, it is easy to see that regardless of whether voter j had y over x, or x over y, or x and y tied, this is an instance of manipulation of a two-element agenda. This proves Claim 2 and completes the proof of Theorem 5.2.1. □

5.3 Another Barberá Theorem

Our second consistency-across-agenda requirement will connect choices from an agenda v to choices made from each two-element subset of v by requiring that the base relation completely determine the social choice function. The adjectives "normal" and "binary" have both been used in this context; we opt for the former terminology.

Definition 5.3.1. A social choice function V is *downward normal* if, for every profile **P** and every agenda v, if $x \in$ V(**P**)(v) then, for every $y \in v$, $x \in$ V(**P**)($\{x, y\}$). V is *upward normal* if, for every profile **P** and every agenda v, if $x \in$ V(**P**)($\{x, y\}$) for every $y \in v$, then $x \in$ V(**P**)(v). V is *normal* if it is both downward normal and upward normal.

The following proposition (whose proof we leave as an exercise) gives an equivalent of normality.

Proposition 5.3.2. *A social choice function V is normal iff it satisfies "acyclic rationality": For every profile* **P**, *there exists a reflexive and complete binary relation R on the set A of alternatives such that R is acyclic (meaning there are no cycles in the strict relation P derived from R) such that, for every agenda* v,

$$\mathrm{x} \in V(\boldsymbol{P})(\mathrm{v}) \text{ iff } \mathrm{x} \in \mathrm{v} \text{ and } \forall \mathrm{y} \in \mathrm{v}, \mathrm{x}R\mathrm{y}.$$

Barberá (1977b) obtained a result that allows us to adapt Corollary 5.1.14 to the context of procedures that are not quasitransitive, but we need to assume, instead, that V is normal, and we need to invoke non-manipulability in the case of a three-element agenda.

Theorem 5.3.3. *In the context of linear ballots, if a social choice function is normal and non-manipulable for two- and three-element agendas, then it is quasitransitive.*

Proof: Notice first that if **P** is a profile in which every voter ranks x over y, then V(**P**)($\{x, y\}$) $= \{x\}$, because otherwise we could choose a profile **P**$'$ such that V(**P**$'$)($\{x, y\}$) $= \{x\}$, and one by one convert the ballots in **P** to those in **P**$'$. At some point we first get an outcome of $\{x\}$ and this contradicts non-manipulability for two-element agendas.

Second, notice that if we have two profiles **P** and **P**$'$ and, for every i, $R_i|\{x, y\} = R'_i|\{x, y\}$, then V(**P**)($\{x, y\}$) $=$ V(**P**$'$)($\{x, y\}$). This is called *binary IIA*. To see that it holds, assume otherwise and convert **P** to **P**$'$ one ballot at a time until the outcome changes. Without loss of generality, assume that $x \in$ V(**P**)($\{x, y\}$), and $x \notin$ V(**P**$'$)($\{x, y\}$). If voter i preferred y to x, then – taking his or her ballot in **P** to be his or her true preferences – he or she has changed the election outcome from $\{x\}$ or $\{x, y\}$ to $\{y\}$. Either way contradicts non-manipulability for two-element agendas. If voter i preferred x to y, then – taking his or her ballot in **P**$'$ to be his or her true preferences – he or she has changed the election outcome from $\{y\}$ to either $\{x, y\}$ or $\{x\}$. Either way again contradicts non-manipulability for two-element agendas.

Now, to prove that V is quasitransitive, assume that $\mathrm{V}(\mathbf{P})(\{x, y\}) = \{x\}$ and $\mathrm{V}(\mathbf{P})(\{y, z\}) = \{y\}$. Because V is downward normal, we know that $\mathrm{V}(\mathbf{P})(\{x, z\}) \neq \{z\}$, or else we would have $\mathrm{V}(\mathbf{P})(\{x, y, z\}) = \emptyset$, and so, if V is not quasitransitive, then $\mathrm{V}(\mathbf{P})(\{x, z\}) = \{x, z\}$. We'll show that this is impossible.

Claim 1. $\mathrm{V}(\mathbf{P})(\{x, y, z\}) = \{x\}$.

Proof. Because $\mathrm{V}(\mathbf{P})(\{x, y\}) = \{x\}$, we know $y \notin \mathrm{V}(\mathbf{P})(\{x, y, z\})$ by downward normality, and because $\mathrm{V}(\mathbf{P})(\{y, z\}) = \{y\}$ we know $z \notin \mathrm{V}(\mathbf{P})(\{x, y, z\})$.

Claim 2. Suppose $\mathbf{P}'$ is the result of some voter moving y to the bottom of his or her ballot. Then $y \notin \mathrm{V}(\mathbf{P}')(\{x, y, z\})$.

Proof. By downward normality, it suffices to show that we still have $\mathrm{V}(\mathbf{P}')(\{x, y\}) = \{x\}$. But if this were not so, then this voter could put y back, and this would be an instance of manipulation of two-element agendas.

Claim 3. If $\mathbf{P}''$ is the result of all voters moving y to the bottom of their ballots, then $\mathrm{V}(\mathbf{P}'')(\{x, y, z\}) = \{x, z\}$.

Proof. We know $y \notin \mathrm{V}(\mathbf{P}'')(\{x, y, z\})$ by Claim 2. Because everyone has x over y, we know, by the first paragraph of this proof, that $x \in \mathrm{V}(\mathbf{P}'')(\{x, y\})$, and, because of binary IIA, $x \in \mathrm{V}(\mathbf{P}'')(\{x, z\})$. Hence $x \in \mathrm{V}(\mathbf{P}'')(\{x, y, z\})$ by upward normality. Similarly, we have that $z \in \mathrm{V}(\mathbf{P}'')(\{x, y, z\})$. This proves Claim 3.

Thus, as we start with the profile $\mathbf{P}$ and, one-by-one, have each voter move y to the bottom of his or her ballot, the election outcome with agenda $\{x, y, z\}$ starts at $\{x\}$ and ends at $\{x, z\}$ with only $\{z\}$ as a possible in-between election outcome.

Case 1. At some point, the election outcome goes from $\{x\}$ to $\{z\}$.

At this point (call the profile $\mathbf{Q}$), upward normality implies that $x \notin \mathrm{V}(\mathbf{Q})(\{x, y\})$ or $x \notin \mathrm{V}(\mathbf{Q})(\{x, z\})$. By binary IIA, it must be the former. Hence $\mathrm{V}(\mathbf{Q})(\{x, y\}) = \{y\}$, and, to arrive at $\mathbf{Q}$, voter i has plunged y to the bottom of his or her ballot. Taking this latter ballot to be his true or her preferences, we have an instance of manipulation of a two-element agenda.

Case 2. At some point, the election outcome goes from $\{x\}$ to $\{x, z\}$.

If voter i had z over x, then the change from $\{x\}$ to $\{x, z\}$ would be an instance of manipulation of a three-element agenda. If voter i had x over z, then the change from $\{x, z\}$ to $\{x\}$ would be an instance of manipulation of a three-element agenda. This completes the proof of Claim 3 and Theorem 5.3.2. □

The assumption in Theorem 5.3.3 that V is non-manipulable on three-element agendas as well as on two-element agendas turns out to be necessary, as can be seen by looking ahead to Example 8.3.1. For now, however, we conclude with a corollary that again characterizes a natural collection of variable agenda social choice functions.

Corollary 5.3.4. *In the context of linear ballots, if* n *is a positive integer and* A *is a set of three or more alternatives, then the following are equivalent for any social choice function for* (A, n)*:*

(1) V is normal, pairwise non-imposed, and non-manipulable on two-and three-element agendas.
(2) V is the "Pareto rule with dummies." There exists a set O of voters (the "non-dummies") such that $\forall \boldsymbol{P}$ *and* $\forall \mathrm{v} \subseteq A$ *and* $\forall \mathrm{x} \in \mathrm{v}$,

$$\mathrm{x} \in V(\boldsymbol{P})(\mathrm{v}) \textit{ iff } \mathrm{x} \in \mathrm{v} \textit{ and } \forall \mathrm{y} \in \mathrm{v}, \mathrm{y} \neq \mathrm{x} \Rightarrow \exists \mathrm{i} \in O \textit{ with } \mathrm{x} P \mathrm{iy}.$$

Notice that (2) is asserting that V|O is (essentially) the Pareto rule on v for every agenda v.

5.4 The MacIntyre–Pattanaik Theorem

Upward normality, we would argue, is not an unreasonable assumption. It is easy to see, for example, that the following social choice function, even with non-linear ballots, is neutral, anonymous, and upward normal, even though no voter can use any alternative to block any other.

Example 5.4.1. The weak Condorcet social choice function C is defined as follows. For any profile **P** and agenda v, C(**P**)(v) is the result of applying the weak Condorcet rule to **P**|v.

With linear ballots and an odd number of voters, the weak Condorcet social choice function is non-manipulable in the sense of weak dominance for two- and three-element agendas, as shown in Theorem 2.2.1 (iii). And, although the weak Condorcet social choice function is not downward normal (Exercise), it does satisfy a weaker version of downward normality given by the following.

Definition 5.4.2. A social choice function V is *weakly downward normal* if, for every profile **P** and every agenda v, if $x \in \mathrm{V}(\mathbf{P})(v)$, then for every $y \in v$, either $x \in \mathrm{V}(\mathbf{P})(\{x, y\})$ or there exists some $z \in v$ with $\mathrm{V}(\mathbf{P})(\{y, z\}) = \{z\}$.

The sequential pairwise voting rule also satisfies weak downward normality (Exercise).

Remarkably, the following theorem of MacIntyre and Pattanaik (1981) shows that with *non-linear* ballots, the only way this can be achieved is with some voters having the power to use one alternative to block another.

Theorem 5.4.3 (The MacIntyre–Pattanaik Theorem for Non-Linear Ballots). *In the context of non-linear ballots, if* n *is a positive integer and A is a set of four or more alternatives, then any social choice function for* (A, n) *that is non-manipulable on two- and three-element agendas, upward normal, and weakly downward normal has a voter who is a nominator on some two-element agenda.*

Before proving Theorem 5.4.3, it is instructive to see why the weak Condorcet social choice function is manipulable (the proof of Theorem 5.4.3 being an elaboration of this). So suppose we have three voters and any number of alternatives. Let **P** be the following profile.

a	c	b
b		c
c	ab	a
.	.	.
.	.	.
.	.	.

If $v = \{a, b, c\}$, then $\mathrm{C}(\mathbf{P})(v) = \{b\}$. But voter 2 prefers the set $\{a, b, c\}$ to the set $\{b\}$ according to the notion of weak dominance. Moreover, he or she can ensure this election outcome by changing his or her ballot to c over a over b.

The proof of Theorem 5.4.5 requires several lemmas.

Lemma 5.4.4 (Weak IIA with Ties in the Ballots). *Suppose* **P** *and* **P**$'$ *are two profiles, and* x *and* y *are two alternatives such that* $P_{\mathrm{j}}|\{\mathrm{x}, \mathrm{y}\} = P'_{\mathrm{j}}|\{\mathrm{x}, \mathrm{y}\}$ *for every* j. *Assume additionally that for every voter* j *who has* x *and* y *tied in* **P** *or* **P**$'$, *we have* $R_{\mathrm{j}} = R'_{\mathrm{j}}$. *Then* $V(\mathbf{P})(\{\mathrm{x}, \mathrm{y}\}) = V(\mathbf{P}')(\{\mathrm{x}, \mathrm{y}\})$.

Proof: Assume $\mathrm{V}(\mathbf{P})(\{x, y\}) \neq \mathrm{V}(\mathbf{P}')(\{x, y\})$, and one by one change the ballots in **P** to those in **P**$'$ until the election outcome for $v = \{x, y\}$ changes as a result of a change in a single ballot, say that of voter i. Now, voter i did not have x and y tied and so he or she preferred x to y on both or y to x on both. Either way represents an instance of manipulation for two-element agendas. □

Lemma 5.4.5 (Monotonicity). *Suppose $\mathbf{P}'$ is a profile arrived at by ballot changes in the profile $\mathbf{P}$, and for every* j, $\mathrm{x}R_{\mathrm{j}}\mathrm{y}$ *implies* $\mathrm{x}R'_{\mathrm{j}}\mathrm{y}$ *and* $\mathrm{x}\mathbf{P}_{\mathrm{j}}\mathrm{y}$ *implies* $\mathrm{x}P'_{\mathrm{j}}\mathrm{y}$. *Then* $V(\mathbf{P})(\{\mathrm{x},\mathrm{y}\}) = \{\mathrm{x}\}$ *implies* $V(\mathbf{P}')(\{\mathrm{x},\mathrm{y}\}) = \{\mathrm{x}\}$, *and* $\mathrm{x} \in V(\mathbf{P})(\{\mathrm{x},\mathrm{y}\})$ *implies* $\mathrm{x} \in V(\mathbf{P}')(\{\mathrm{x},\mathrm{y}\})$.

Proof: If not, we could do the ballot changes one by one, and, at the point at which the change occurred, regard the preferences in $\mathbf{P}'$ to be the true preferences. This would be an instance of manipulation of two-element agendas. □

Lemma 5.4.6 (The Existence Lemma). *Suppose that for some voter* i *and some alternatives* x *and* y, *there exists a profile* $\mathbf{P}$ *such that* $\mathrm{x}P_{\mathrm{i}}\mathrm{y}$, *and for every* $\mathrm{j} \neq \mathrm{i}$, $\mathrm{y}P_{\mathrm{j}}\mathrm{x}$. *Assume* $\mathrm{x} \in V(\mathbf{P})(\{\mathrm{x},\mathrm{y}\})$. *Then voter* i *can use* x *to block* y.

Proof: Suppose $\mathbf{P}'$ is any profile in which voter i has $x\mathrm{P}'_i y$ and assume $\mathrm{V}(\mathbf{P}')(\{x, y\}) = \{y\}$. By Lemma 5.4.5, everyone with x and y tied in $\mathbf{P}'$ can move y above x and we still have $\{y\}$ as the winner. The same is true if everyone but voter i who had x over y interchanges x and y. But we now have two profiles $\mathbf{P}$ and $\mathbf{P}'$ such that $\mathrm{P}_j|\{x, y\} = \mathrm{P}'_j|\{x, y\}$ for every j (and in which no voter has x and y tied). It thus follows from Lemma 5.4.4 that $\mathrm{V}(\mathbf{P})(\{x, y\}) = \mathrm{V}(\mathbf{P}')(\{x, y\})$, a contradiction. □

Now, if X is a set of voters and (x, y) is an ordered pair of distinct alternatives, we'll say that X is a *dictating set for the pair* (x, y) if, for every profile $\mathbf{P}$, $\mathrm{V}(\mathbf{P})(\{x, y\}) = \{x\}$ whenever $x\mathrm{P}_i y$ for ever $i \in \mathrm{X}$. Notice that by Lemma 5.4.5, we lose no generality in assuming that $y\mathrm{P}_j x$ for every $j \notin \mathrm{X}$. Notice also that voter i can use x to block y iff $\mathrm{N} - \{i\}$ fails to be a dictating set for the pair (x, y). In particular, if there are no dictating sets for any pairs, then the proof is complete.

Choose a set X such that X is a dictating set for some pair (a, b) and choose X such that no smaller set is a dictating set for any pair of alternatives. If $|\mathrm{X}| = 1$, then we are done, because that single voter in X would then be able to use x to block y in quite a strong sense.

Assume, then, that $|X| > 1$, and choose distinct voters $i, j \in X$ and distinct alternatives $c, d \in A - \{a, b\}$. Let **P** be the following profile:

i	j	$X - \{i, j\}$	$N - X$
d	c	a	b
a	ab	b	d
b		c	c
c	d	d	a

Case 1. $b \in V(\mathbf{P})(\{a, b\})$; i.e., bRa, where R is the base relation. Consider $v = \{a, b, c\}$. Then $\mathbf{P}|v$ is the following:

i	j	$X - \{i, j\}$	$N - X$
a	c	a	b
b	ab	b	c
c		c	a

Because i's ballot agrees with everyone in $X - \{i, j\}$ on $\{a, b, c\}$, we can condense this as follows:

j	$X - \{j\}$	$N - X$
c	a	b
ab	b	c
	c	a

Everyone but voter j has b over c, so either j can use c to block b (in which case we are done) or bPc. We also have cRa, or else $X - \{j\}$ would contradict the choice of X being of minimal size. Hence, we can apply our assumptions of normality for the agenda $v = \{a, b, c\}$, knowing that $bPcRa$ and bRa.

Because V is upward normal, we know $b \in V(\mathbf{P})(\{a, b, c\})$, and because V is weakly downward normal, $c \notin V(\mathbf{P})(\{a, b, c\})$. Hence, $V(\mathbf{P})(\{a, b, c\}) = \{b\}$ or $\{a, b\}$. Let $\mathbf{P}'$ be the profile in which voter j breaks his or her ab tie by placing a over b.

Case 1.1 $c \in V(\mathbf{P}')(\{a, b, c\})$.

In this case, voter j has gained by the manipulation, because he or she prefers any set containing c to $\{b\}$ or $\{a, b\}$. But notice that this is no longer a comparison of two-element sets.

Case 1.2 $V(\mathbf{P}')(\{a, b, c\}) = \{a\}$.

In this case, we can have the voters in $N - X$ one by one change their ballots to c over b over a. By IIA, we still have $cRaPb$. But now we also have

$c\text{R}b$, because $\text{X} - \{j\}$ is not a dictating set for the pair (b, c). Thus, by upward normality, some voter in $\text{N} - \text{X}$ has changed his or her ballot so that the winner went from his or her singleton last choice to something else. This is an instance of manipulation.

Case 1.3 $\text{V}(\mathbf{P}')(\{a, b, c\}) = \{b\}$.

In this case, we can have the voters in $\text{N} - \text{X}$ one by one change their ballots to b over a over c. The winner now becomes $\{a\}$ and undoing this last change is an instance of manipulation.

Case 1.4 $\text{V}(\mathbf{P}')(\{a, b, c\}) = \{a, b\}$.

In this case, we can again have the voters in $\text{N} - \text{X}$ one by one change their ballots to b over a over c. If the first time the election outcome changes results in $\{b\}$, then this is clearly an instance of manipulation. If the result is $\{a\}$, $\{c\}$ or $\{a, c\}$, then undoing this change is an instance of manipulation. If the winner becomes $\{b, c\}$ or $\{a, b, c\}$, then the normality assumptions yield $a\text{P}b\text{P}c\text{P}a$. We can now put voter j's ballot back at c over ab-tied and still have $b\text{P}c\text{P}a$ (by IIA), and $b\text{R}a$. Thus b is a winner and c is a non-winner, so the outcome is $\{b\}$ or $\{a, b\}$. Either way, j can undo the tie and get a better outcome, so this is an instance of manipulation.

Case 2. $\text{V}(\mathbf{P})(\{a, b\}) = \{a\}$; i.e., $a\text{P}b$, where R is the base relation.

Consider $v = \{a, b, d\}$. Then $\mathbf{P}|v$ is the following:

i	j	$\text{X} - \{i, j\}$	$\text{N} - \text{X}$
d	ab	a	b
a	d	b	d
b		d	a

Notice that $\text{X} - \{i\}$ has a over d, and everyone else has d over a. Because $\text{X} - \{i\}$ is not a dictating set for (a, d), we must have $d\text{R}a$. Moreover, we have everyone except voter i has b over d and so, unless voter i can use d to block b, we have $b\text{P}d$. With the assumption of Case 2, this yields $a\text{P}b\text{P}d\text{R}a$.

Case 2.1. $d \in \text{V}(\mathbf{P})(\{a, b, d\})$.

Let voter j break his tie so that he has b over a over d. By IIA, we still have $b\text{P}d\text{R}a$, but we now also have $b\text{R}a$ or else $\text{X} - \{j\}$ would be a dictating set for the pair (a, b). Thus, b is a winner and d is a loser, and this represents an instance of manipulation by voter j.

It now follows that either $\text{V}(\mathbf{P})(\{a, b, d\}) = \{a\}$, $\{b\}$ or $\{a, b\}$. The arguments here now follow quite closely those for Cases 1.2, 1.3, and 1.4. We leave the details for the reader. □

6
Ultrafilters and the Infinite

6.1 The Infinite Version of Arrow's Theorem

In many ways, Arrow's theorem is mathematically more natural in the context of infinitely many voters than it is in the context of finitely many voters. For example, with infinitely many voters, one is led to the conclusion of Arrow's theorem with no feeling of impending paradox, the disquieting nature of the result only revealing itself in the corollary pertaining to the finite.

Throughout this section, we work with an arbitrary, perhaps infinite, set N of voters and an arbitrary, again perhaps infinite, set A of three or more alternatives. Our focus is on the social welfare version of Arrow's theorem, and we assume for simplicity that we are in the context of linear ballots.

Our starting point is Condorcet's original intuition that aggregation procedures with three or more alternatives should take advantage of the fact that voting with two alternatives presents few problems. Condorcet's voting rule (wherein an alternative is a unique winner iff it can defeat every other alternative in a one-on-one contest) is certainly inspired by this intuition. Moreover, with social welfare functions, there is an obvious kinship between this intuition and the desire to have the property of independence of irrelevant alternatives satisfied.

So let us begin with an imprecise description – the imprecision located by the quotation marks – of a social welfare function based on Condorcet's intuition and see where it leads us.

(*) For any linear profile **P** and alternatives $a, b \in A$, let's agree to rank a over b in the final list V(**P**) iff a "large set of voters" rank a over b on their ballots in **P**.

There are three notions of "large set of voters" that immediately suggest themselves. First, if N is finite, we might want to interpret "large set of voters" to mean "more than half the voters." This, however, does not yield a social welfare function. Indeed, as Condorcet himself observed, if three voters have ballots *abc*, *bca*, and *cab*, then the final list would have a strictly preferred to b and b strictly preferred to c. But c is strictly preferred to a, and this shows that the "social ordering" is not even quasitransitive.

A second thought is to have "large set of voters" interpreted as meaning "includes every voter" – so N itself is the only large set. This approach is superior to what we just tried in that we now avoid cycles in the strict preferences of the final list. However, if two voters have ballots *abc* and *cab*, then the final list would have b weakly preferred to c and c weakly preferred to a. But b is not weakly preferred to a, because both voters have a over b, and this shows that the "social ordering," although quasitransitive, is not transitive.

Indeed, we know from Arrow's theorem that if N is finite, then the only notion of "large set of voters" that will yield a transitive and complete social ordering (and have the set N itself large) is one obtained by fixing a particular voter – a dictator – and declaring the "large sets" to be precisely the ones to which the dictator belongs. This is true because Pareto would follow from the fact that N is assumed to be "large" and IIA is a consequence of the binary nature of the definition.

To be more precise, suppose we start with a collection U of subsets of N, with the sets in U being the ones that we intuitively want to think of as "large." Consider the function $\mathrm{F_U}$ that associates to each N-sequence **P** of A-ballots (hereafter called an (A, N)-profile) the binary relation $\mathrm{F_U}(\mathbf{P})$ given by

$$(a, b) \in \mathrm{F_U}(\mathbf{P}) \text{ iff } \{i \in \mathrm{N}\colon (a, b) \in \mathrm{P}_i\} \in \mathrm{U}.$$

The function $\mathrm{F_U}$ so defined will be called a *social welfare relation* to distinguish it from a social welfare function, which, as we've seen, it might well fail to be. The question of which social welfare relations are social welfare functions is answered by the following.

Theorem 6.1.1. *The social welfare relation F_U is a social welfare function satisfying Pareto and IIA iff the collection U of subsets of N satisfies the following three properties:*

(1) $N \in U$.
(2) If a set in U is partitioned into two disjoint sets, then exactly one of the sets must also be in U.
(3) If $X \in U$, and $X \subseteq Y \subseteq N$, then $Y \in U$.

Proof: ($\Rightarrow$) Suppose that the social welfare relation $\mathrm{F_U}$ is a social welfare function satisfying Pareto and IIA. Because $\mathrm{F_U}$ satisfies Pareto, we have $\mathrm{N} \in \mathrm{U}$. Now suppose that the set $\mathrm{X} \in \mathrm{U}$ is partitioned into disjoint sets Y and Z. If neither

Y nor Z is in U, then we get a contradiction by considering the profile in which the voters in Y have ballots yzx, the voters in Z have ballots zxy, and the voters in N − X have ballots xyz. If both Y and Z are in U, then we get a contradiction by considering the profile in which the voters in Y have ballots xyz, the voters in Z have ballots zxy, and the voters in N − X have ballots xzy. Finally, suppose that $X \in U$ and $X \subseteq Y \subseteq N$. If Y is not in U, we get a contradiction by considering the profile in which the voters in X have ballots yxz, the voters in Y − X have ballots yzx, and the voters in N − Y have ballots zyx.

($\Leftarrow$) We need only verify that $F_U(\mathbf{P})$ is always a linear ordering of the set A. So suppose that **P** is an arbitrary (A, N)-profile and that a and b are two distinct elements of A. Let $X = \{i \in N: (a, b) \in P_i\}$ and let $Y = N - X$. Since $N \in U$ by (1), we know by (2) that at least one of X and Y is in U, establishing completeness, and that at most one of X and Y is in U, establishing antisymmetry. For transitivity, we need only establish that if $X \in U$ and $Y \in U$, then $X \cap Y \in U$. But if this were not true, then (2) would yield $X - Y \in U$ and $Y - X \in U$, which, using (3), contradicts (2). This completes the proof. □

Remark. The independence of the three conditions can be seen by taking N to be a three-element set and letting U be (respectively) the empty collection, the collection containing only N itself, and the collection containing N itself together with the singleton subsets of N.

By definition, a collection U of subsets of N is an *ultrafilter on N* if it satisfies conditions (1), (2), and (3) in Theorem 6.1.1.[22] For each $i \in N$, the collection $U_i = \{X \subseteq N: i \in X\}$ is an ultrafilter on N, and these are called *principal ultrafilters*. If N is finite, then *every* ultrafilter on N is principal (Exercise). This observation allows us to conclude – on the basis of Theorem 6.1.1 – that a social welfare relation of the form F_U for a finite set N of voters is a social welfare function satisfying Pareto and IIA iff it is a dictatorship. But every *neutral* social welfare function is of the form F_U (Exercise). Thus, Theorem 6.1.1 proves Arrow's theorem for the special case in which the social welfare function is neutral.

[22] The usual definition (see Comfort and Negrepontis, 1974, p. 143 or Bell and Slomson, 1969) goes as follows: A non-empty collection F of subsets of N is a *filter on N* if (1) $\emptyset \notin F$; (2) if X, $Y \in F$, then $X \cap Y \in F$; and (3) if $X \in F$ and $X \subseteq Y \subseteq N$, then $Y \in F$. The filter U on N is an *ultrafilter* on N if it is not properly contained in any other filter on N (and this happens iff for every $X \subseteq N$, at least one of X and N − X is in U). This definition is equivalent to what we gave above (Exercise).

The full version of Arrow's theorem – for an arbitrary set of voters – is an immediate consequence of Theorem 6.1.1 and the following (which should be attributed to Arrow).

Theorem 6.1.2. *Every social welfare function that satisfies Pareto and IIA is neutral.*

Proof: Assume that F satisfies Pareto and IIA. For each $(a, b) \in A \times A$, IIA ensures that whether or not we have $(a, b) \in F(\mathbf{P})$ depends only on $\{i \in N: (a, b) \in P_i\}$. Hence, there is a collection U_{ab} of subsets of N such that $(a, b) \in F(\mathbf{P})$ iff $\{i \in N: (a, b) \in P_i\} \in U_{ab}$.[23] To establish neutrality, we must show that $U_{ab} = U_{xy}$ for any two distinct elements x and y of A.

We first note that if $X \in U_{ab}$ and $c \neq a$, then $X \in U_{ac}$ as can be seen by considering the profile in which the voters in X have abc and the others have bca. Similarly, if $X \in U_{ab}$, then $X \in U_{cb}$, as can be seen by considering the profile in which the voters in X have cab and the others have bca.

It now follows that if $X \in U_{yx}$ then $X \in U_{xy}$ (by way of $X \in U_{yc}$ and $X \in U_{xc}$), so we're done if $a = y$ and if $b = x$. If $a \neq y$ then $X \in U_{ab}$ implies $X \in U_{ay}$ which implies $X \in U_{xy}$, and we're done. If $a = y$ and $b \neq x$, then $X \in U_{ab}$ is $X \in U_{yb}$, which implies $X \in U_{yx}$, and so we get $X \in U_{xy}$ as desired. This completes the proof. □

Putting Theorems 6.1.1 and 6.1.2 together yields a result that should properly be attributed to Hansson (1976) and Kirman and Sondermann (1972) although their work was preceded by earlier observations of Blau, unpublished, Fishburn (1970) and Guilbaud (1952). It is the general (ultrafilter) version of Arrow's theorem for an arbitrary set of voters and an arbitrary set of three or more alternatives. For other treatments, see Armstrong (1980) and Lauwers and Van Liedekerke (1995).

Theorem 6.1.3 (Hansson, Kirman, Sondermann, Blau, Fishburn, Arrow). *With an arbitrary set N of voters and an arbitrary set A of three or more alternatives, a social welfare function for A and N satisfies Pareto and IIA iff it is of the form F_U for some ultrafilter U on N, where*

$$(a, b) \in F_U(\mathbf{P}) \text{ iff } \{i \in N: (a, b) \in P_i\} \in U.$$

Kirman and Sondermann (1972) referred to ultrafilters as "invisible dictators." We borrow this name for the title of the next two sections.

[23] The cognoscenti will recognize that X is decisive for a against b iff every superset of X is in U_{ab}.

6.2 Infinite Gibbard–Satterthwaite without Invisible Dictators

An observation dating back at least to Pazner and Wesley (1977 and 1978) is that if the set A of alternatives is finite, then every ultrafilter U on a (perhaps infinite) set N also induces a resolute voting rule V_U as follows:

$$V_U(\mathbf{P}) = x \text{ iff } \{i \in N\colon \max(P_i) = x\} \in U.$$

Moreover, it is easy to see that such a resolute voting rule satisfies Pareto and is non-manipulable. But – unlike Arrow's theorem – the Gibbard–Satterthwaite theorem does not directly generalize to this context. In particular, the analog of Theorem 6.1.2 is false; neutrality does *not* follow from non-manipulability and Pareto when the set N of voters is infinite. The example in the following proof is very close to one in Mihara (2000).

Theorem 6.2.1. *If N is infinite and the set A of alternatives is finite, then there exist resolute voting rules that are non-manipulable, satisfy Pareto, and are not neutral (and thus not induced by any ultrafilter on N).*

Proof: Let $<a_1, a_2, \ldots, a_k>$ be an enumeration of the elements in A, and let V be defined by $V(\mathbf{P}) = a_j$ iff j is the least integer for which infinitely many voters rank a_j at the top of their ballot. The fact that V satisfies Pareto but not neutrality is trivial. The fact that V is non-manipulable by any individual is also trivial, but an interesting realization of the political scientist's "paradox of voting": No one's individual vote matters. This completes the proof. □

A critical aspect of Theorem 6.2.1 is the single-voter nature of the manipulation. In point of fact, the voting rule in that proof *can* be manipulated by a coalition in the following sense. Suppose that every voter in the infinite set X ranks a_3 first, a_1 second, and a_2 third, and suppose that N − X is also infinite and that every voter in N − X ranks a_2 first. Then a_2 wins. But every voter in X prefers a_1 to a_2, and the voters in X, acting together (with the ballots of those voters not in X held fixed), can arrange for a_1 to win simply by moving it to the top of their ballots. This kind of "coalitional manipulability" is, in the case where N is finite and Pareto holds, equivalent to single-voter manipulability.[24] Theorem 6.2.1 shows that this equivalence fails when N is infinite.

[24] Clearly, single-voter manipulability implies coalitional manipulability because a single voter is a coalition of size one. However, single-voter non-manipulability implies (in the presence of Pareto) that the resolute voting rule is a dictatorship, and dictatorships are coalitionally non-manipulable.

6.3 Invisible Dictators Resurrected

In this section, we give a new proof of a result having roots in Pazner and Wesley (1977), Batteau, Blin, and Monjardet (1981), and, more recently, Mihara (2000). This result asserts that if we replace individual manipulability by coalitional manipulability, then we can obtain an ultrafilter version of the Gibbard–Satterthwaite theorem for the case in which the set of alternatives is finite. We do this by deriving it as a corollary to Arrow's theorem. This requires passing from a resolute voting rule to a social welfare function. In Section 6.4, we prove a stronger result by yet another proof. We continue to work with a perhaps infinite set N of voters, and to use (A, N) in place of our earlier (A, n).

Assume, then, that V is a resolute voting rule for (A, N) that satisfies Pareto. The claim is that V gives rise to a social welfare function F_V for (A, N) in a natural way that can be described informally as follows. Given an (A, N)-profile **P**, we begin by letting the top element of the social ordering $F_V(\mathbf{P})$ be the election winner V(**P**). To get the second element of the social ordering, we move the election winner V(**P**) to the bottom of everyone's ballot, and then take the second element of the social ordering to be the new winner (and because V satisfies Pareto, it is indeed new). Iterating this yields the rest of the social ordering.

We need a slightly more precise description. If P_i is a linear (A, N)-ballot and $x \in A$, then we let xP_i be the ballot arrived at by moving x to the bottom of P_i. We can now define the social ordering $F_V(\mathbf{P})$ to be the sequence $<a_1, \ldots, a_k>$ where $a_1 = V(\mathbf{P})$ and, for $2 \leq j \leq k$, $a_j = V(<a_{j-1} \ldots a_1 P_i>)$.

Theorem 6.3.1. *If V is a resolute voting rule that satisfies Pareto and is coalitionally non-manipulable, then the induced social welfare function F_V satisfies Pareto and IIA.*

Proof: Suppose that F_V fails to satisfy Pareto and choose **P** such that $F(\mathbf{P}) = <a_1, \ldots, a_k>$ and, for some $s < t$, we have $(a_t, a_s) \in P_i$ for every i. Let $Q_i = a_{s-1} \ldots a_1 P_i$. Then $V(\mathbf{Q}) = a_s$. But $a_t \notin \{a_1, \ldots, a_{s-1}\}$ because $s < t$ and so, in passing from **P** to **Q**, no voter moved a_t down. Thus we have $(a_t, a_s) \in Q_i$ for every i, showing that V also fails to satisfy Pareto.

Now suppose that F_V fails to satisfy IIA; we'll show that some coalition can manipulate V. We first claim that there exist alternatives $a, b \in A$ and profiles **P** and **P**′ such that $P_i|\{a, b\} = P'_i|\{a, b\}$ for every $i \in N$, yet $V(\mathbf{P}) = a$ and $V(\mathbf{P}') = b$. To see this, choose alternatives $a, b \in A$ and profiles **Q** and **Q**′ such that $Q_i|\{a, b\} = Q'_i|\{a, b\}$ for every $i \in N$, but $(a, b) \in F(\mathbf{Q})$ and $(b, a) \in F(\mathbf{Q}')$. For notation, let $F(\mathbf{Q}) = <a_1, \ldots, a_k>$ with $a = a_r$ and let $F(\mathbf{Q}') = <b_1, \ldots, b_k>$, with $b = b_s$. Let $P_i = a_{r-1} \ldots a_1 Q_i$ and let $P'_i = b_{s-1} \ldots b_1 Q'_i$. The only changes in passing from Q_i to P_i and Q_i to Q'_i involve moving alternatives distinct from a and b to the bottom of ballots. Thus, $P_i|\{a, b\} = Q_i|\{a, b\} = Q'_i|\{a, b\} = P'_i|\{a, b\}$ for every $i \in N$, and $V(\mathbf{P}) = a$ and $V(\mathbf{P}') = b$.

With this choice of a, b, $\mathbf{P}$, and $\mathbf{P}'$, let X and Y be as follows:

$$X = \{i \in N: (a, b) \in P_i\} = \{i \in N: (a, b) \in P'_i\}$$
$$Y = \{i \in N: (b, a) \in P_i\} = \{i \in N: (b, a) \in P'_i\}.$$

Let $A - \{a, b\} = \{c_3, \ldots, c_k\}$, and let P_{ab} and P_{ba} be the following ballots:

$$P_{ab} = <a, b, c_3, \ldots, c_k> \text{ and } P_{ba} = <b, a, c_3, \ldots, c_k>.$$

Now consider the following three profiles:

$$R_i = \begin{cases} P_{ab} \text{ if } i \in X \\ P_i \text{ if } i \in Y \end{cases} \qquad R'_i = \begin{cases} P'_i \text{ if } i \in X \\ P_{ba} \text{ if } i \in Y \end{cases} \qquad S_i = \begin{cases} P_{ab} \text{ if } i \in X \\ P_{ba} \text{ if } i \in Y \end{cases}$$

By Pareto, $V(\mathbf{S}) = a$ or $V(\mathbf{S}) = b$. Without loss of generality, assume that it is the latter. If $V(\mathbf{R}) \neq a$, then everyone in X, with R_i as their true preferences, could switch from R_i to P_i and get their first choice a as the election outcome. This is coalitional manipulability. If $V(\mathbf{R}) = a$, then everyone in Y, with R_i as their true preferences, could switch from R_i to P_{ba} and get b, which they all prefer to a, as the election outcome. This again is coalitional manipulability, and completes the proof. □

We conclude with the desired consequence of Theorem 6.1.3 and Theorem 6.3.1.

Theorem 6.3.2. *With an arbitrary set N of voters and a finite set A of three or more alternatives, a resolute voting rule for A and N satisfies Pareto and coalitional non-manipulability iff it is of the form V_U for some ultrafilter U on N where*

$$V_U(\boldsymbol{P}) = \mathrm{x} \textit{ iff } \{\mathrm{i} \in N: max(P_\mathrm{i}) = \mathrm{x}\} \in U.$$

Pareto can be replaced in Theorem 6.3.2 by non-imposition (Exercise). For related work, see Mihara (2001).

6.4 Infinitely Many Voters and Infinitely Many Alternatives

The difference between the known infinite version of Arrow's theorem (Theorem 6.1.3) and the known infinite version of the Gibbard–Satterthwaite theorem (Theorem 6.3.2) is that the former deals with an arbitrary set of alternatives and the latter requires that the set of alternatives be finite. In what follows we address this by generalizing Theorem 6.3.2 via techniques quite different from Mihara (2000).

With infinitely many voters and alternatives, ultrafilters provide natural (if you will) examples of social welfare functions defined for every (A, N)-profile. But the same is not true for resolute voting rules, and with good reason. For example, no resolute voting rule can satisfy Pareto unless we restrict the class of profiles to which it applies. To see this, let P be an A-ballot with no largest element and let **P** be the profile in which each voter has P as his or her ballot.

Even for profiles in which all ballots have a maximal element, ultrafilters seem to be of no help in constructing resolute voting rules unless they are principal (in which case the resolute voting rule is a dictatorship). On the other hand, each ultrafilter on N does give rise to a natural collection of (A, N)-profiles and to a resolute voting rule on this class that is reminiscent of Condorcet's method and that turns out to satisfy Pareto and coalitional non-manipulability. Moreover, this class of profiles is general enough so that if A is finite, no profiles are omitted, and thus the result it leads to is a strict generalization of Theorem 6.3.2.

We cannot, however, be too cavalier in restricting the domain of resolute voting rules. The point is that our proof will involve invoking Arrow's theorem in the case of social welfare functions, and any restriction in the domain of the resolute voting rule will carry over to a restricted domain for the induced social welfare function. Thus, we need to make sure that the class of profiles to which our voting rule applies is general enough to allow the proof of Arrow's theorem to go through.

Definition 6.4.1. A non-empty collection $\Im$ of (A, N)-profiles is said to be an *admissible class* if $\mathbf{Q} \in \Im$ whenever there exists a profile $\mathbf{P} \in \Im$ and a finite set $B \subseteq A$ such that $P_i|(A - B) = Q_i|(A - B)$ for every $i \in N$.

If A is finite, then there are no proper admissible subclasses of the set of all (A, N)-profiles. If A is infinite, however, there are many; for example, the set of (A, N)-profiles made up of ballots that have an initial segment ordered as $a_1 > a_2 > \ldots$ Other examples of admissible classes arise from ultrafilters as we show momentarily.

The proof we gave of the general version of Arrow's theorem (Theorem 6.1.3) involved moving around only finitely many alternatives at a time. Thus, we have the following generalization of that result.

Theorem 6.4.2. *Assume that A is an arbitrary set of three or more alternatives, N is an arbitrary set of voters, and $\Im$ is an admissible class of (A, N)-profiles. Assume that F is a social welfare function for $\Im$ and N that satisfies Pareto and IIA. Then there exists an ultrafilter U on N such that, for every profile $\mathbf{P} \in \Im$,*

$$(\mathrm{a}, \mathrm{b}) \in F(\mathbf{P}) \text{ iff } \{\mathrm{i} \in N\colon (\mathrm{a}, \mathrm{b}) \in P_{\mathrm{i}}\} \in U.$$

Each ultrafilter on N gives rise to a natural collection of (A, N)-profiles that turns out to be an admissible class, and to a resolute voting rule on this class that is reminiscent of Condorcet's voting rule and that turns out to satisfy Pareto and coalitional non-manipulability. The description of the class and method runs as follows.

Definition 6.4.3. If U is any collection of subsets of N, then an alternative a is a *U-Condorcet winner (UCW) for the profile* $\mathbf{P}$ if, for every alternative $b \neq a$,

$$\{i \in \mathrm{N}\colon (a, b) \in \mathrm{P}_i\} \in \mathrm{U}.$$

We give three examples:

(1) If A is finite, and $\mathrm{U} = \{\mathrm{X} \subseteq \mathrm{N}\colon |\mathrm{X}| > |\mathrm{A}|/2\}$, then a is a U-Condorcet winner for $\mathbf{P}$ iff a is a Condorcet winner in the usual sense.
(2) If A is finite, and $\mathrm{U} = \{\mathrm{X} \subseteq \mathrm{N}\colon |\mathrm{X}| \geq |\mathrm{A}|/2\}$, then a is a U-Condorcet winner for $\mathbf{P}$ iff a is a weak-Condorcet winner in the usual sense.
(3) If $\mathrm{U} = \{\mathrm{X} \subseteq \mathrm{N}\colon \mathrm{j} \in \mathrm{X}\}$ is a principal ultrafilter on N, then a is a U-Condorcet winner for $\mathbf{P}$ iff a is at the top of voter j's ballot.

Now, if U is an ultrafilter on N, then the collection of (A, N)-profiles for which there exists a U-Condorcet winner is an admissible class that we denote by $\Im_{\mathrm{U}}$. Moreover, we can define a resolute voting rule V_{U} on $\Im_{\mathrm{U}}$ by setting $\mathrm{V}_{\mathrm{U}}(\mathbf{P}) = a$ iff a is the (necessarily unique) UCW for $\mathbf{P}$. Note that if $\mathrm{U} = \{\mathrm{X} \subseteq \mathrm{N}\colon j \in \mathrm{X}\}$ is a principal ultrafilter on N, then $\Im_{\mathrm{U}}$ is the collection of profiles in which voter j has a top-most element on his ballot, and V_{U} is the dictatorship in which voter j is the dictator.

It is straightforward to verify that the resolute voting rules V_U all satisfy Pareto and are coalitionally non-manipulable. But much more is true.

Theorem 6.4.4. *Assume that A is an arbitrary (perhaps infinite) set of three or more alternatives, N is an arbitrary (perhaps infinite) set of voters, and $\Im$ is an admissible class of (A, N)-profiles. Assume that V is a resolute voting rule for $\Im$ and N that satisfies Pareto and is coalitionally non-manipulable, Then there exists an ultrafilter U on N such that, for every profile $\mathbf{P} \in \Im$,*

$$V(\mathbf{P}) = \text{a } \textit{iff } \text{a } \textit{is the (necessarily unique) UCW for } \mathbf{P}.$$

Proof: Given V, let F_V be the social welfare relation for $(\Im, N)$ given by:

$$(a, b) \in F_V(<P_i>) \text{ iff } V(<abP_i>) = a.$$

We will show that F_v is a social welfare function that satisfies Pareto and IIA, and that, moreover, $V(\mathbf{P}) = a$ iff for every $b \in A$, $V(<abP_i>) = a$. This will suffice, because we then know, by Theorem 6.4.2, that F_V is of the form F_U for some ultrafilter U on N. With this (and the verification that we will provide that $V(<P_i>) = a$ iff for every $b \in A$, $V(<abP_i>) = a$) we have $V(<P_i>) = a$ iff for every $b \neq a$, $V(<abP_i>) = a$ iff for every $b \neq a$, $(a, b) \in F_V(<P_i>)$ iff for every $b \neq a$, $(a, b) \in F_U(<P_i>)$ iff for every $b \neq a$ $\{i \in N: (a, b) \in P_i\} \in U$ iff $V_U(<P_i>) = a$.

To see that $F_V(<P_i>)$ is complete and antisymmetric, note that if a and b are distinct, then $(a, b) \in F_V(<P_i>)$ iff $V(<abP_i>) = a$ iff $V(<baP_i>) = a$ (because $abP_i = baP_i$) iff $V(<baP_i>) \neq b$ (by Pareto) iff $(b, a) \notin F_V(<P_i>)$. Assume now that $F_V(<P_i>)$ is not transitive, and choose $a, b, c \in A$ such that $(a, b) \in F_V(<P_i>)$, $(b, c) \in F_V(<P_i>)$, and $(a, c) \notin F_V(<P_i>)$. By the completeness of $F_V(<P_i>)$, we have $(c, a) \in F_V(<P_i>)$. Thus, $V(<abP_i>) = a$, $V(<bcP_i>) = b$, and $V(<caP_i>) = c$. By Pareto, $V(<abcP_i>) \in \{a, b, c\}$, and so we can assume, with no loss of generality, that $V(<abcP_i>) = b$. Let $X = \{i \in N: (a, b) \in P_i\}$, $Y = \{i \in N: (b, a) \in P_i\}$, and let $<Q_i>$ be the profile wherein $Q_i = abcP_i$ if $i \in X$ and $Q_i = abP_i$ if $i \in Y$.

Suppose first that $V(<Q_i>) = b$. Consider the voters in X with true preferences given by Q_i. Every voter in this coalition prefers a to b, but $V(<Q_i>) = b$. However, if these voters in X simultaneously change from $abcP_i$ to abP_i, then every voter in N has abP_i, and $V(<abP_i>) = a$. Thus X has engaged in successful coalitional manipulability.

Suppose now that $V(<Q_i>) \neq b$. Consider the voters in Y with true preferences given by Q_i. Every voter in this coalition has b at the top of his or her ballot, but $V(<Q_i>) \neq b$. However, if these voters in Y simultaneously change from $ab\text{P}_i$ to $abc\text{P}_i$, then every voter has $abc\text{P}_i$, and $\text{V}(<abc\text{P}_i>) = b$. Thus Y has engaged in successful coalitional manipulability. This shows that $\text{F}_V(<\text{P}_i>)$ is always transitive, and thus that F_V is a social welfare function.

For Pareto, note that if $(a, b) \in \text{P}_i$ for every $i \in \text{N}$, then $\text{V}(<ab\text{P}_i>) = a$ because V satisfies Pareto, and thus $(a, b) \in \text{F}_V(<\text{P}_i>)$. Assume now that IIA fails, and choose $<\text{P}_i>$ and $<\text{Q}_i>$ such that $\text{P}_i|\{a, b\} = \text{Q}_i|\{a, b\}$ for every $i \in \text{N}$, $(a, b) \in \text{F}_V(<\text{P}_i>)$, and $(b, a) \in \text{F}_V(<\text{Q}_i>)$. Thus, $\text{V}(<ab\text{P}_i>) = a$ and $\text{V}(<ab\text{Q}_i>) = b$. Let $\text{X} = \{i \in \text{N}: (a, b) \in \text{P}_i\} = \{i \in \text{N}: (a, b) \in \text{Q}_i\}$, let $\text{Y} = \{i \in \text{N}: (b, a) \in \text{P}_i\} = \{i \in \text{N}: (b, a) \in \text{Q}_i\}$, and let $<\text{S}_i>$ be the profile wherein $\text{S}_i = ab\text{P}_i$ if $i \in \text{X}$ and $\text{Q}_i = ab\text{Q}_i$ if $i \in \text{Y}$. Note that $\text{V}(<\text{S}_i>) \in \{a, b\}$ by Pareto.

Suppose first that $\text{V}(<\text{S}_i>) = a$. Consider the voters in X with true preferences given by $ab\text{Q}_i$. Every voter in this coalition prefers a to b, but $\text{V}(<ab\text{Q}_i>) = b$. However, if these voters in X simultaneously change from $ab\text{Q}_i$ to $ab\text{P}_i$, then the profile $<ab\text{Q}_i>$ changes to $<\text{S}_i>$ and $\text{V}(<\text{S}_i>) = a$. Thus X has engaged in successful coalitional manipulability.

Suppose now that $\text{V}(<\text{S}_i>) = b$. Consider the voters in Y with true preferences given by $ab\text{P}_i$. Every voter in this coalition prefers b to a, but $\text{V}(<ab\text{P}_i>) = a$. However, if these voters in Y simultaneously change from $ab\text{P}_i$ to $ab\text{Q}_i$, then the profile $<ab\text{P}_i>$ changes to $<\text{S}_i>$ and $\text{V}(<\text{S}_i>) = b$. Thus Y has engaged in successful coalitional manipulability. This shows that $\text{F}_\text{V}(<\text{P}_i>)$ satisfies Pareto and IIA.

For the left-to-right direction of the moreover-clause, assume that $\text{V}(<\text{P}_i>) = a$ but $\text{V}(<ab\text{P}_i>) \neq a$ for some b. By Pareto, $\text{V}(<ab\text{P}_i>) = b$. Let $\text{X} = \{i \in \text{N}: (a, b) \in \text{P}_i\} = \{i \in \text{N}: (a, b) \in ab\text{P}_i\}$, let $\text{Y} = \{i \in \text{N}: (b, a) \in \text{P}_i\} = \{i \in \text{N}: (b, a) \in ab\text{P}_i\}$, and let $<\text{S}_i>$ be the profile wherein $\text{S}_i = ab\text{P}_i$ if $i \in \text{X}$ and $\text{S}_i = \text{P}_i$ if $i \in \text{Y}$.

Suppose first that $\text{V}(<\text{S}_i>) \neq a$. Consider the voters in X with true preferences given by $<\text{S}_i>$. Every voter in this coalition has a at the top of his or her ballot, but $\text{V}(<\text{S}_i>) \neq a$. However, if these voters in X simultaneously change from $ab\text{P}_i$ to P_i, then the profile $<\text{S}_i>$ changes to $<\text{P}_i>$ and $\text{V}(<\text{P}_i>) = a$. Thus X has engaged in successful coalitional manipulability.

Suppose now that $\text{V}(<\text{S}_i>) = a$. Consider the voters in Y with true preferences given by S_i. Every voter in this coalition prefers b to a, but $\text{V}(<\text{S}_i>) = a$. However, if these voters in Y simultaneously change from S_i to $ab\text{P}_i$, then the profile $<\text{S}_i>$ changes to $<ab\text{P}_i>$ and $\text{V}(<ab\text{P}_i>) = b$. Thus Y has engaged in successful coalitional manipulability.

Finally, for the right-to-left direction of this clause, assume that $\text{V}(<\text{P}_i>) \neq a$, and choose b such that $\text{V}(<\text{P}_i>) = b$. Then, by the left-to-right direction of the moreover clause, we have $V(<baP_i>) = b$, and so $V(<abP_i>) \neq a$. This completes the proof. □

Of course, if the set A of alternatives is finite, then we have the known result that there exists an ultrafilter U on N such that for every **P**,

$$V(\mathbf{P}) = a \text{ iff } \{i \in N\text{: } \max(P_i) = a\} \in U.$$

PART THREE

7

More on Resolute Procedures

7.1 Combinatorial Equivalents

Our proof of the Gibbard–Satterthwaite theorem in Section 3.1 was based on the combinatorial property that we called down-monotonicity, and the fact (Lemma 3.1.5) that if a resolute voting rule is non-manipulable, then it satisfies down-monotonicity. It turns out that the converse of this is also true, so down-monotonicity is fully equivalent to non-manipulability for resolute voting rules.

Notice, however, that we are not assuming Pareto, so the equivalence of down-monotonicity and non-manipulability does not follow from the implications "non-manipulability implies down-monotonicity implies dictatorship implies non-manipulability." Results in the next chapter, nevertheless, allow a version of this derivation.

Down-monotonicity is not the only combinatorial equivalent of non-manipulability for resolute voting rules. In what follows, we give three others, two of which have appeared in the literature (Muller and Satterthwaite, 1977, and Moulin, 1983), and the third of which is the "resolute version" of the property we called "choice independence of irrelevant alternatives" earlier.

Down-monotonicity is a particularly appealing combinatorial property because it comes in two versions that are easily seen to be equivalent. One of these versions is quite weak: An election winner is unchanged if one voter moves one losing alternative down one spot on his or her ballot. The other version is quite strong: An election winner is unchanged if several voters move several losing alternatives down several spots on their ballots. In what follows, we make use of these two versions of down-monotonicity to show that three other combinatorial properties are equivalent to non-manipulability. Throughout this section, V denotes a resolute voting rule, and we continue to work only in the context of linear ballots.

Definition 7.1.1. In the context of linear ballots, suppose that V is a resolute voting rule for (A, n). Then V satisfies:

(1) *strong monotonicity* provided that the following holds for every election: If an alternative is moved up one spot on a voter's ballot, then either the winner is unchanged, or that alternative becomes the new winner.
(2) *strong positive association* provided that the following holds for every election: If a is the winner, and ballots are changed but no voter moves an alternative he or she had ranked below a to a position above a, then a remains the winner.
(3) *choice independence of irrelevant alternatives* provided that the following holds for every election: If a is the winner and b is some other alternative, and ballots are changed but no voter reverses the relative rankings of a and b on his or her ballot, then b remains a non-winner.

Recall that a resolute voting rule in the context of linear ballots is *monotone* if the winner of an election is unchanged when any voter exchanges the position of the winning alternative on his or her ballot with that of the alternative directly above it on his or her ballot. We can now state the desired equivalencies.

Theorem 7.1.2. *In the context of linear ballots, if V is a resolute voting rule for (A,* n*), then the following are equivalent:*

(1) V is non-manipulable.
(2) V satisfies down-monotonicity.
(3) V satisfies strong-monotonicity.
(4) V satisfies strong positive association.
(5) V satisfies montonicity and choice independence of irrelevant alternatives.

Proof: One usually proves a result such as this by a sequence of five implications, beginning with (1) implies (2) and ending with (5) implies (1). It is, however, more illuminating in this case to take advantage of the fact that each of the other five properties is easily seen to imply the weak version of down-montonicity, and to follow from the strong version of down-monotonicity. The implications are represented by a double arrow in the following list:

(1) $\Rightarrow$ (2): This is Lemma 3.1.5.

(2) $\Rightarrow$ (1): Assume V satisfies down-monotonicity, and suppose, for contradiction, that V is manipulable. Then we have two elections, in the first of which y is the winner even though voter i has ranked x over y, and in the second of which x is the winner, with only voter i having changed his or her ballot (in some unspecified way). Using the strong version of down-monotonicity, we

can have voter i start plunging all alternatives other than x and y, in some fixed order, to the bottom of his or her ballot. If voter i does this in both elections, then his or her ballots will wind up identical in the two elections, except that he or she will have x first and y second in the first election (where y was the winner), and y first and x second in the second election (where x was the winner). But now in the first election, he or she can move x just below y (using down-monotonicity on the losing alternative x) and have identical ballots in the two elections, a contradiction.

(2) $\Rightarrow$ (3): Assume strong monotonicity fails. Then we have two elections, in the first of which voter i has x immediately over y on his or her ballot, and the winner is some alternative v, and in the second of which he or she has interchanged the positions of x and y on his or her ballot, with the election result becoming some w that is neither v nor y. But now, starting with the second election, we can see that down-monotonicity fails by moving the losing alternative y (losing because $w \neq y$) down one slot on voter i's ballot and seeing that the winner changes from w to v (a change because $w \neq v$).

(3) $\Rightarrow$ (2): Assume down-monotonicity fails. Then we have two elections, in the first of which voter i has x immediately over y on his or her ballot, and the winner is some alternative $v \neq x$, and in the second of which he or she has interchanged the positions of x and y on his or her ballot (moving the losing alternative x down one spot), with the election result becoming some $w \neq v$. But now, starting with the second election, we can see that strong monotonicity fails by moving x up one spot on voter i's ballot and seeing that the winner changes from w to v but with $v \neq x$.

(2) $\Rightarrow$ (4): Assume V satisfies down-monotonicity, and fix a voter i and a sequence of ballots in which alternative a is the winner. Because all of the alternatives that voter i has ranked above a are non-winners, we can achieve any permutation of these by using down-monotonicity. Similarly, we can achieve any permutation of the alternatives ranked below a on voter i's ballot. This shows that strong positive association holds.

(4) $\Rightarrow$ (2): Assume down-monotonicity fails. Then we have two elections, in the first of which voter i has x immediately over y on his or her ballot, and the winner is some alternative $v \neq x$, and in the second of which he or she has interchanged the positions of x and y on his or her ballot (moving the losing alternative x down one spot), with the election result becoming some $w \neq v$. But clearly voter i did not move any alternative that was below v to above v because $v \neq y$. Nevertheless, the election winner changed, and this shows that strong positive association fails.

(2) $\Rightarrow$ (5): Because V satisfies down-monotonicity, V is easily seen to satisfy monotonicity. Now assume, for contradiction, that choice independence

of irrelevant alternatives fails. Then we have two elections in which no voter changes the relative position of a and b, but a is the winner of the first, and b is the winner of the second. Because of down-monotonicity, we can assume that all the alternatives distinct from a and b are ranked below a and b on every ballot, and in the same fixed order. But then the ballots in the two elections are identical, and yet the winners are not, and this is a contradiction.

(5) $\Rightarrow$ (2): Assume down-monotonicity fails. Then we have two elections, in the first of which voter i has x immediately over y on his or her ballot, and the winner is some alternative $v \neq x$, and in the second of which he or she has interchanged the positions of x and y on his or her ballot (moving the losing alternative x down one spot), with the election result now becoming some $w \neq v$. This is a violation of choice independence of irrelevant alternatives, unless $\{v, w\} = \{x, y\}$. However, we also know that $v \neq y$, so not only are the two sets equal, but, in fact, $v = x$ and $w = y$. This means that y wins the first election (in which voter i had x over y) and x wins the second election (in which voter i had y over x). This is a violation of monotonicity. □

We conclude this section with two comments. First, we could, in fact, define "choice independence of irrelevant alternatives for single voters" in the obvious way, and this would be another property equivalent to the five properties in Theorem 7.1.1. Second, in the context of voting rules that are not necessarily resolute, even the five equivalencies in Theorem 7.1.1 no longer all hold.

7.2 Characterization Theorems for Resolute Voting Rules

The primary characterization theorem that we seek in this section answers the following question: Without assuming Pareto or the existence of three or more distinct alternatives, can we still identify exactly which resolute voting rules – in the context of linear ballots – are non-manipulable? Our starting point, of necessity, is with the case in which we have only two alternatives.

Thus, for the moment, we fix a two-element set $A = \{x, y\}$ as our set of alternatives. The goal is to identify which resolute voting rules on A are non-manipulable. We begin with a definition from combinatorics.

Definition 7.2.1. A *hypergraph* G on N is a pair G = (N, E) where N is an arbitrary set and E is a collection of subsets of N. Elements of N are called *vertices* and sets in E are called *edges*. If every set in E has exactly two elements, then G is a *graph*.

The characteristic function of a hypergraph is a *Boolean function*, and much of what we know about hypergraphs has been provided by Boolean function theorists over the last half century or so.[25] But the relevance of hypergraphs or Boolean functions to our present discussion is given by the following correspondence.

To each resolute voting rule V on the two-element set A = $\{x, y\}$ of alternatives, we can associate two hypergraphs:

(1) $G_{V,x} = (N, E_{V,x})$ where $Z \in E_{V,x}$ iff the election winner is x when the voters in Z are precisely the ones who rank x over y on their ballots.
(2) $G_{V,y} = (N, E_{V,y})$ where $Z \in E_{V,y}$ iff the election winner is y when the voters in Z are precisely the ones who rank y over x on their ballots.

For example, if V is the voting rule in which x is the winner regardless of what the ballots look like, then $G_{V,x} = (N, \wp(N))$ and $G_{V,y} = (N, \emptyset)$.

In an analogous fashion, we can associate to each hypergraph G = (N, E), two resolute voting rules on the two-element set A = $\{x, y\}$ of alternatives:

(1) $V_{G,x}$, in which x is the winner if the set Z of voters who rank x over y is in E.
(2) $V_{G,y}$, in which y is the winner if the set Z of voters who rank y over x is in E.

For example, if we start with the hypergraph $G = (N, \wp(N))$, then $V_{G,x}$ is the voting rule in which x is the winner regardless of what the ballots look like, and $V_{G,y}$ is the voting rule in which y is the winner regardless of what the ballots look like.

It turns out that there is a well-known connection between the two hypergraphs that arise from a single resolute voting rule on A = $\{x, y\}$. This connection is given by the following definition.

Definition 7.2.2. If G = (N, E) is a hypergraph, then the *dual of G* is the hypergraph $G^d = (N, E^d)$, where $X \in E^d$ iff $N - X \notin E$. The hypergraph G is *constant sum* if $G = G^d$.

It is easy to see that $G^{dd} = G$, and so G is the dual of H iff H is the dual of G. In particular, if we define a relation $\equiv$ on hypergraphs by setting $G \equiv H$ iff $G = H^d$ or $G = H$, then $\equiv$ is an equivalence relation. Notice that each equivalence class with respect to $\equiv$ consists of either one hypergraph (and this happens exactly when the hypergraph is constant sum) or two hypergraphs, each of which is the dual of the other.

[25] For more on this, see the references in Taylor and Zwicker (1999).

In an analogous fashion, we can define the dual of a resolute voting rule V on the set $A = \{x, y\}$.

Definition 7.2.3. With a two-element set $A = \{x, y\}$ of alternatives, the *dual of V* is the resolute voting rule V^d defined so that $V^d(P_1, \ldots, P_n) = x$ iff $V({P_1}^d, \ldots, P_n^d) = y$, where P_i^d is the ballot that has x over y iff P_i has y over x. V is *neutral* if $V = V^d$.

Again it is easy to see that $V^{dd} = V$, and so V is the dual of V′ iff V′ is the dual of V. In particular, if we define a relation $\equiv$ on resolute voting rules by setting $V \equiv V'$ iff $V' = V^d$ or $V = V'$, then $\equiv$ is an equivalence relation. Notice that each equivalence class with respect to $\equiv$ consists of either one voting rule (and this happens exactly when the voting rule is neutral) or two voting rules, each of which is the dual of the other. We can summarize the preceding discussion in the following proposition.

Proposition 7.2.4. *There is a natural one-to-one correspondence between equivalence classes of hypergraphs on the set N and equivalence classes of resolute voting rules with voter set N and having a two-element set of alternatives. Under this correspondence, the constant-sum hypergraphs correspond to the resolute voting rules that are neutral.*

Given this identification, one can now ask the following: If we consider the collection of non-manipulable resolute voting rules, is the corresponding collection of hypergraphs a natural class in its own right? The answer is yes, and it requires the following definition.

Definition 7.2.5. A *simple game* G on N is a hypergraph (N, E) that is *monotone* (or *monotonic*): If $X \in E$ and $X \subseteq Y \subseteq N$, then $Y \in E$.

Game theorists view games in a number of different ways, one of which involves a function that assigns real numbers to sets. Simple games correspond to the special case in which the values of such a function are restricted to be zero or one.

Simple games arise in a number of different contexts, including, for example, reliability theory, wherein elements of N are identified with components of a system, and sets in E with collections of components sufficient for the operation of the system as a whole. For more on this, see the text by Ramamurthy (1990).

But perhaps the most natural context for simple games is a voting-theoretic setting that is different from the one with which we are presently dealing, and which involves a legislative body that votes on issues that are placed before it. Each such issue is pitted against the status quo, and each voter gets to weigh in with his or her approval or disapproval. A "yes–no voting system" is a set of rules that determines exactly which collections of voters can ensure passage of the issue at hand by voting in favor of it. Simple games provide a natural model of such voting systems; for more on this, see the texts by Taylor (1995) and Taylor and Zwicker (1999).

The proof of the following proposition is quite trivial (and omitted) given the observation that a hypergraph $G = (N, E)$ is monotone iff $X \cup \{x\} \in E$ whenever $X \in E$ and $x \in N$.

Proposition 7.2.6. *With two alternatives, a resolute voting rule is non-manipulable iff the hypergraph corresponding to it is a simple game.*

At this point, we are ready to return to the general case in which we have more than two alternatives. Let's say that an alternative x is *viable* if there exists at least one profile **P** for which $V(\mathbf{P}) = x$. Of course, if there are three or more alternatives and they are all viable, then we know that the only non-manipulable resolute voting rules are the dictatorships (and that dictatorships are, indeed, non-manipulable resolute voting rules).

In general, however, some of the alternatives might not be viable. Nevertheless, it is relatively easy, at least in the case of linear ballots, to completely characterize the non-manipulable resolute voting rules.

Theorem 7.2.7. *In the context of linear ballots, a resolute voting rule is non-manipulable iff one of the following holds:*

(1) There is an ordered pair (x, y) *of (not necessarily distinct) alternatives, and a simple game G on N such that the winner is* x *iff the set of voters who rank* x *over* y *is an edge in G, and the winner is* y *otherwise. Notice that if* $y = x$ *and* $G = (N, \wp(N))$*, then* x *is the winner regardless of the ballots.*
(2) There is a set B containing three or more of the alternatives, and a particular voter i *such that the winner of an election is the alternative in B that is ranked highest by voter* i.

Proof: It is easy to see that all the voting rules given in (1) and (2) are resolute and non-manipulable. For the converse, we assume that V is resolute and non-manipulable, and we let B be the set of alternatives in A that are viable.

The key to making use of Theorem 3.1.2 and the propositions established earlier in this section is the following observation. Let $\mathbf{P} = \langle P_1, \ldots, P_n \rangle$ be an arbitrary (A, n)-profile, and define the resolute voting rule $V_{\mathbf{P}}$ on B by setting $V_{\mathbf{P}}(S_1, \ldots, S_n) = V(T_1, \ldots, T_n)$ where T_i agrees with S_i on B and with P_i on $A - B$ for every i. Now let $\mathbf{P}' = \langle P'_1, \ldots, P'_n \rangle$ be any other (A, n)-profile.

Claim: $V_{\mathbf{P}} = V_{\mathbf{P}'}$.

Proof. Suppose not. This means that for some (B, n)-profile $<S_1, \ldots, S_n>$, we have $V_{\mathbf{P}}(S_1, \ldots, S_n) \neq V'_{\mathbf{P}}(S_1, \ldots, S_n)$. Thus, $V(T_1, \ldots, T_n) \neq V(T'_1, \ldots, T'_n)$ where both T_i and T'_i agree with S_i on B for every i. Now, since only the alternatives in B are viable for V, this means that the winner switched from being some $x \in B$ to some $y \in B$ even though no one changed the relative order of any alternatives in B on his or her ballot. But it is easy to see that such a change caused by a single voter is an instance of manipulation. This contradiction proves the claim.

It thus follows that we can consider an induced resolute voting ruleV^* on B defined by applying V to the profile obtained by adding the alternatives in A – B, in any positions we desire, to each of the ballots. It is also easy to see that V^* is non-manipulable (because V is), and that every alternative in B is viable for V^* as well.

Now, if $|B| = 1$ or $|B| = 2$, then the results earlier in this section yield the desired simple game G. If $|B| \geq 3$, then Theorem 3.1.2 yields the voter i as in (2). This completes the proof of Theorem 7.2.7. □

7.3 Characterization Theorems for Resolute Social Choice Functions

As pointed out earlier, a social choice function acquires coherence only in the presence of some kind of consistency property such as quasitransitivity (giving consistency of choices among agendas of the same size) or normality (giving consistency of choices among agendas of different sizes). Without the imposition of such consistency properties, one could simply choose, for each agenda v, a social choice function V_v and set $V(\mathbf{P})(v) = V_v(\mathbf{P}|v)$, where $\mathbf{P}|v = (\mathbf{P}_1 \cap (v \times v), \ldots, P_n \cap (v \times v))$. For example, $V_{\{a,b,c\}}$ might be the Borda count and $V_{\{x,y,z\}}$ might be the Hare system.

But not every social choice function can be arrived at in this way. In fact, the ones that can be so described are precisely the ones that satisfy independence of irrelevant alternatives (IIA), as set forth in Theorem 1.3.1: For every two profiles $\mathbf{P}$ and $\mathbf{P}'$ and every agenda v, if $R_i|v = R'_i|v$ for every i, then $V(\mathbf{P})(v) = V(\mathbf{P}')(v)$.

The benefit of considering resolute variable agenda social choice functions is twofold:

(1) The definition of manipulability (Definition 7.3.1 below) is uncontroversial, as all of the difficulties in comparing sets of alternatives evaporate.
(2) It is easy to prove (Proposition 7.3.2 below) that V satisfies IIA and so all of the results from Chapters 2 and 3 apply to $V|v$ for every agenda v.

In the remainder of this section we provide the aforementioned definition and proposition, and then make use of earlier results to precisely characterize those resolute social choice functions that are non-imposed, nonmanipulable, and normal. For the rest of this section, V denotes a resolute social choice function.

Definition 7.3.1. V is *manipulable* if there exists an agenda v and a profile $\mathbf{P} = (P_1, \ldots, P_n)$, which we think of giving the true preferences of the n voters, and another ballot Q_i, which we think of as a disingenuous ballot from voter i such that, letting $\mathbf{P}' = (P_1, \ldots, P_{i-1}, Q_i, P_{i+1}, \ldots, P_n)$, we have

$$V(\mathbf{P}')(v) P_i V(\mathbf{P})(v).$$

Proposition 7.3.2. *If V is non-manipulable, then V satisfies IIA.*

Proof: Assume that $\mathbf{P}$ and $\mathbf{P}'$ are two profiles such that $R_i|v = R_i'|v$ for every i. If $V(\mathbf{P})(v) \neq V(\mathbf{P}')(v)$, then we can one by one convert the ballots in $\mathbf{P}$ to those in $\mathbf{P}'$ until a change in one voter's ballot – call him or her voter i – causes the change in election outcome from, say, x to y. But x and y are both in v, and so they are ranked the same way by voter i in $\mathbf{P}$ and in $\mathbf{P}'$. If he or she ranks x over y, we can take voter i's ballot in $\mathbf{P}'$ to be his or her true preferences, and if he or she ranks y over x, we can take voter i's ballot in $\mathbf{P}$ to be his or her true preferences. Either way yields an instance of manipulation. □

At this point we can invoke any of the results from Chapters 2 and 3 that we wish to. But we will be content to state just the one result alluded to above.

Theorem 7.3.3. *V is a dictatorship iff V is non-imposed, non-manipulable, and normal.*

Proof: It is trivial to verify that a dictatorship satisfies the three conditions. For the converse, consider the resolute social choice function V′ defined on all of A by $V'(\mathbf{P}) = V(\mathbf{P})(A)$. Because V′ is non-imposed and non-manipulable (and because we are assuming that $|A| > 2$), we know there is a voter i who is a dictator for V′.

Now suppose that $v = \{x, y\}$ and assume that $\mathbf{P}$ is a profile in which $x P_i y$ and assume, for the moment, that $V(\mathbf{P})(\{x, y\}) = y$. Because V satisfies IIA, we can assume that voter i ranks x first on his or her ballot and y second. But now, because V is normal, we know that, because $x \notin V(\mathbf{P})(\{x, y\})$, we have $x \notin V(\mathbf{P})(A)$, Thus, $x \notin V(\mathbf{P}')$, even though the dictator (voter i) has x on top of his or her ballot. This contradiction shows that for every two-element agenda, the winner is whichever alternative voter i has on top of his or her ballot. But it now follows from normality that, for every agenda v and every profile $\mathbf{P}$, $\mathrm{V}(\mathbf{P})(v) = x$ iff x is the element of v ranked highest on his or her ballot. □

7.4 Characterizations for Resolute Social Welfare Functions

Our goal in the present section is to show how the dictatorship-like consequences of non-manipulability can coexist with a quasidemocratic ability of each voter to have a unilateral effect on the outcome of an election, although not necessarily on an equal basis with every other voter. Our approach is to shift contexts from resolute voting rules, where the outcome of an election is a single alternative, to (resolute) social welfare functions, where the outcome of an election is a linear ordering of the set of alternatives – what we will call a "final list." We are still assuming that ballots are linear, and we are likewise disallowing ties in the final list produced by a social welfare function.

Suppose then that L is a linear ordering of the alternatives and that this linear ordering represents a voter's true preferences. Suppose also that L_1 and L_2 are two linear orderings that can arise in the final list using some social welfare function. What would it mean to say that this voter prefers one of the lists to the other?

The answer that we work with is based on a natural notion of manipulability arising from a consideration of lexicographic orderings: Given a list L, which we think of as a ballot giving a voter's true preferences, and two other lists L_1 and L_2, which we think of as possible election outcomes according to some social welfare function, we scan down the two lists until we reach the first place they differ. At this point, we see which alternative is better according to the preference list L.

It turns out that non-manipulability here is (with slight hedging) equivalent to the system's being one in which some voter gets to pick which alternative will be in first place, another voter then gets to specify which will be in second place, and so on, allowing one voter to play more than one role. But first, we need a bit of precision.

Definition 7.4.1. In the context of linear ballots, a resolute social welfare function V is *manipulable* if there exists a profile $\mathbf{P} = (P_1, \ldots, \mathrm{P}_n)$, which we think of as giving the true preferences of the n voters, and another ballot Q_i, which we think of as a disingenuous ballot from voter i such that, letting

$\mathbf{P}' = (P_1, \ldots, P_{i-1}, Q_i, P_{i+1}, \ldots, P_n)$, we have:

(1) $V(\mathbf{P}) = < x_1 \ldots x_p a \ldots >$,
(2) $V(\mathbf{P}') = < x_1 \ldots x_p b \ldots >$, and
(3) $bP_i a$ – that is, b is ranked above a on the ballot P_i giving voter i's true preferences.

Although dictatorships (wherein we fix one of the voters and the final list is simply taken to be his or her ballot) are certainly non-manipulable in this sense, there are more interesting examples, of which we consider two. For the first, assume that voter 1 gets to specify which alternative is on top of the final list, then voter 2 gets to specify which of the remaining alternatives is second, then voter 3 gets a similar say, and so on. Of course, we could modify this by returning to voter 1 for a decision as to which alternative, for example, is third on the final list.

But the example that best illustrates the general case is the following. Suppose we have three voters and three alternatives: a, b, and c. Voter 1 gets to choose which of the three alternatives is at the top of the final list. If voter 1 chooses a, then voter 2 gets to choose which alternative is second on the final list. On the other hand, if voter 1 chooses b, then voter 3 gets to choose which alternative is second. But, if voter 1 chooses c, then which of the remaining two alternatives is second on the final list is determined by majority vote, based on the ballots cast.

There are two ways in which the latter example is more complicated than the former. First, the alternatives are not all treated the same in the latter example – the question of which voter will pick what is second on the final list depends on which alternative voter 1 has on top of his or her ballot. Second, the order in which the bottom two alternatives in the final list appear in the latter example is not decided by a particular voter, but by majority rule. This has no effect on the question of manipulability – in spite of the Gibbard–Satterthwaite theorem – because we are dealing with only *two* alternatives at this point in the process.

As to the first complication, we will opt for simplicity in what follows and consider only those social welfare functions that treat all alternatives the same – that is, those that are "neutral" in the sense of Definition 7.4.2 below. We do, however, briefly comment on the non-neutral case immediately following Theorem 7.4.3.

Definition 7.4.2. In the context of linear ballots, a resolute social welfare function V is *neutral* if, for every permutation σ of the set A of alternatives, we have

$$V(\sigma(P_1), \ldots, \sigma(P_n)) = \sigma(V(P_1, \ldots, P_n))$$

where, for every linear ordering $L = <a_1, \ldots, a_k>$ of A, $\sigma(L)$ is defined to be $<\sigma(a_1), \ldots, \sigma(a_k)>$.

With these preliminaries at hand, we can now state the result derivable from the Gibbard–Satterthwaite theorem that shows how non-manipulability can coexist with the kind of quasidemocracy that we alluded to above.

Theorem 7.4.3. *In the context of linear ballots, suppose we have a set* $A = \{x_1, \ldots, x_k\}$ *of two or more alternatives and a resolute social welfare function V that is neutral. Then the following are equivalent:*

(1) V is non-manipulable and for every linear ordering L of the set of alternatives, if $\mathbf{P} = <L, \ldots, L>$, *then* $V(\mathbf{P}) = L$.

(2) Either:

(i) There exists a sequence (repetitions allowed) $<i_1, \ldots, i_{k-1}>$ *such that for* $p \leq k - 1$, *the* p^{th} *alternative* x_p *on the final list* $<x_1, \ldots, x_k>$ *is the alternative in* $A - \{x_1, \ldots, x_{p-1}\}$ *ranked highest by voter* i_p, *or*

(ii) there exists a sequence (repetitions allowed) $<i_1, \ldots, i_{k-2}>$ *and a constant-sum simple game* $G = <N,W>$ *such that, for* $p \leq k - 2$, *the* p^{th} *alternative* x_p *on the final list* $<x_1, \ldots, x_k>$ *is the alternative in* $A - \{x_1, \ldots, x_{p-1}\}$ *ranked highest by voter* i_p, *and the order of the last two alternatives in the final list is determined by the constant-sum simple game G in the sense that* $\{i\text{: voter } i \text{ ranks } x_{k-1} \text{ over } x_k\} \in W$.

Proof: Clearly (2) implies (1); we derive the converse from the Gibbard–Satterthwaite theorem. Given a social welfare function V, we begin by inductively constructing a sequence $<i_1, \ldots, i_{k-2}>$ of voters. To obtain i_1, consider the voting rule V′ obtained by setting $V'(\mathbf{P})$ equal to the top alternative on the list $V(\mathbf{P})$. If $|A| \geq 3$, then V′ satisfies the hypotheses of the Gibbard–Satterthwaite theorem. Hence, there is a voter i_1 such that the top alternative on his or her ballot is always the top alternative on the list $V(\mathbf{P})$.

Now, fix one alternative a and let V″ be the voting rule for the set $A - \{a\}$ of alternatives that is defined as follows. If **P** is a profile for $A - \{a\}$, then let **P′** be the profile for A obtained by placing a at the top of all ballots in **P**. Now let $V''(\mathbf{P})$ be the second alternative on the list $V(\mathbf{P}')$. Note that $V''(\mathbf{P}) \neq a$, because a is the top alternative in the list $V(\mathbf{P}')$ because voter i_1 has a on top of his or her ballot.

Again, if $|A - \{a\}| \geq 3$, we claim that V″ satisfies the hypotheses of the Gibbard–Satterthwaite theorem. To see that it is non-manipulable, assume that the profile **P** represents the true preferences over $A - \{a\}$ of the voters, that

$V''(\mathbf{P}) = b$, that **Q** is a profile that results from a change by voter j alone, and that $V''(\mathbf{Q}) = c$, where c is preferred to b on voter j's ballot in **P**. Let $\mathbf{P}'$ and $\mathbf{Q}'$ be obtained by placing a at the top of all the ballots. Then voter j can change the outcome with V from $<ab\ldots>$ to $<ac\ldots>$, and he or she will prefer the second list to the first according to our lexicographic definition.

Hence, there is a voter i_2 such that, if all voters have alternative a on top of their ballots, then the second alternative of the final list is the alternative that he or she (voter i_2) ranks highest among those in $A - \{a\}$. In the general case, i_2 will be a function of a, as in our earlier three-voter, three-alternative example. But with our assumption of neutrality, we must have i_2 independent of which alternative voter i_1 has on top of his or her ballot.

We now claim that if **P** is a profile in which voter i_1 has a on top, then voter i_2's top-ranked alternative in $A - \{a\}$ is in second place on the final list $V(\mathbf{P})$ regardless of where any of the other voters (except voter i_1) place a. To see this, suppose $\mathbf{P}'$ is a profile showing otherwise; thus, voter i_2 has b as the highest ranked alternative in $A - \{a\}$, but the outcome is a list $ac\ldots$ with $c \neq b$. One by one move a to the top of each ballot in $\mathbf{P}'$ until the final list changes so that it begins $ax\ldots$ with $x \neq c$ (and there must be such a point because it is true when everyone has moved a to the top). But the last voter to make this change had x and c ranked the same way on both ballots, and so he or she has succeeded in manipulating the outcome according to our lexicographic definition.

We can now consider two fixed alternatives a and b and repeat the argument above with ballots for $A - \{a, b\}$, and an election outcome being the third-ranked alternative in the final list arrived at by placing a first and b second on all these ballots. This yields voter i_3, and we can continue this process until we have only two alternatives left. At this point, the Gibbard–Satterthwaite theorem no longer applies, but it is easy to see that we can obtain the constant-sum simple game G by saying that a set X is in W iff the final ordering of these two alternatives agrees with the way they are ordered by voters in X whenever all the voters in X have them ordered one way and everyone else has them ordered the opposite way. A special case of this will be when a set is winning iff it contains some voter i_k. This special case gives us conclusion (i) instead of conclusion (ii). This completes the proof of the theorem. □

Without the assumption of neutrality, clause (2i) would have to be modified so that i_p would depend on the sequence $<x_1, \ldots, x_{p-1}>$, and clause (2ii) would need to be further modified to read: There exists a sequence (repetitions allowed) $<i_1, \ldots, i_{k-2}>$, an ordering $<a_1, \ldots, a_k>$ of the alternatives, and, for each $1 \leq i < j \leq k$, a simple game $G_{ij} = <\{1, \ldots, n\}, W_{ij}>$ such that, for p = 1, . . . , $k - 2$, if V(**P**) = $<x_1, \ldots, x_k>$, then x_p is the alternative in A − $\{x_1, \ldots, x_{p-1}\}$) ranked highest by voter i_p, and $\{x_{k-1}, x_k\} = \{a_i, a_j\}$ with $i < j$ and either $x_{k-1} = a_1$ and

$$\{\text{r: voter } r \text{ ranks } a_i \text{ over } a_j\} \in W_{ij},$$

or $x_k = a_1$ and

$$\{\text{r: voter } r \text{ ranks } a_i \text{ over } a_j\} \in W_{ij}.$$

8
More on Non-Resolute Procedures

8.1 Gärdenfors' Theorem

In Chapter 4 we showed that all of the non-imposed voting rules that have no nominator can be manipulated by either an optimist or a pessimist. This result, however, does not extend to yield weak-dominance manipulability of such voting rules. Nevertheless, there is a significant collection of non-resolute voting rules that are manipulable in this stronger sense of weak dominance.

Definition 8.1.1. Given a profile $\mathbf{P} = (R_1, \ldots, R_n)$ of not necessarily linear ballots, x is a *weak Condorcet winner* (WCW) iff

$$\forall y \in A, |\{i: xR_i y\}| \geq |\{i: yR_i x\}| .$$

A voting rule V satisfies the *weak Condorcet winner criterion* (WCWC) if, for every profile **P**, the set V(**P**) of winners is precisely the set of WCWs whenever this set is non-empty.

Our starting point is the following:

Theorem 8.1.2. *Suppose A is a set of three or more alternatives,* $n \geq 3$ *and V is a voting rule for* (A, n) *that satisfies the WCWC.*

(1) In the context of non-linear ballots, V is weak-dominance manipulable.
(2) In the context of linear ballots, V is weak-dominance manipulable if the number n *of voters is even, but not necessarily so if* n *is odd.*

Proof: For (1), let **P** be profile in which voters 1, 2, and 3 rank the three alternatives x, y, and z as follows:

y	x	xz
z	y	
x	z	y

and in which voters 1, 2, and 3 have all other alternatives below these and tied with each other. Assume that all other voters have all the alternatives tied. Then it is easy to see that x is a WCW and that y and z are not. Hence, $\mathrm{V}(\mathbf{P}) = \{x\}$.

Now let $\mathbf{P}'$ be the result of voter 1 interchanging y and z on his or her ballot. Then x and z are WCWs and y is not, so $\mathrm{V}(\mathbf{P}') = \{x, z\}$. Thus, voter 1, with true preferences in $\mathbf{P}$, has improved the outcome in his or her eyes by going from $\{x\}$ to $\{x, z\}$. This shows that V is weak-dominance manipulable.

For (2), suppose first that the number of voters is even, and that $\mathrm{A} = \{x, y, z, w_1, \ldots, w_n\}$. Let $\mathbf{P}$ be the profile in which the first four voters have the following ballots:

z	x	y	x
y	z	x	z
x	y	z	y
w_1	w_1	w_1	w_1
.	.	.	.
.	.	.	.
.	.	.	.
w_n	w_n	w_n	w_n

Half the remaining voters rank the alternatives $x, y, z, w_1, \ldots, w_n$ and the other half rank them just the opposite (this requires an even number of voters).

Notice that $\mathrm{V}(\mathbf{P}) = \{x\}$ because x is the only WCW. Now let $\mathbf{P}'$ be the result of voter 1 interchanging y and z on his or her ballot. Then x and y are the only WCWs and so $\mathrm{V}(\mathbf{P}') = \{x, y\}$. Thus, voter 1, with true preferences in $\mathbf{P}$, has improved the outcome in his or her eyes by going from $\{x\}$ to $\{x, y\}$. This shows that V is weak-dominance manipulable.

Finally, if n is odd, we can let V be the voting rule in which the set of winners is the set of WCWs if this set is non-empty, and A otherwise – i.e. the weak-Condorcet rule from Chapter 1. Notice that because ballots are linear and the number of voters is odd, a WCW, if it exists, is unique – i.e., V coincides the Condorcet voting rule from Chapter 1. But Theorem 2.2.1(iii) showed that the Condorcet rule is never weak-dominance manipulable. This completes the proof of Theorem 8.1.2. □

Although Theorem 8.1.2 has some merit, the WCWC is really quite strong. A more reasonable notion (introduced in Chapter 1 but recalled here for convenience) is the following:

Definition 8.1.3. Given a profile $\mathbf{P} = (R_1, \ldots, R_n)$ of not necessarily linear ballots, an alternative x is a *Condorcet winner* (CW) if

$$\forall y \in A, |\{i\colon xR_i y\}| > |\{i\colon yR_i x\}| .$$

A voting rule satisfies the *Condorcet winner criterion* (CWC) if, for every profile **P**, the set V(**P**) contains only the unique CW whenever a CW exists.

With linear ballots and an odd number of voters, an alternative x is a CW iff it is a WCW. However, with linear ballots and an odd number of voters, Theorem 8.1.2 shows that there are voting rules that are not weak-dominance manipulable. This raises the question of whether or not there is a slightly weaker notion of manipulability that is implied by the CWC.

The answer is provided by a 1976 theorem of Peter Gärdenfors, published in the *Journal of Economic Theory*. This theorem involves a weaker notion of manipulability based on what he calls a "sure-thing principle" (and which we take slight exception to in the following discussion). This principle, according to Gärdenfors (1976, p. 220), asserts that a voter will consider a change in election outcome from X to Y to be favorable provided that,

> if some alternative has been added [in passing from X to Y], it should be at least as good as all the other alternatives [in X], and if some alternative has been deleted, it should be worse than the remaining alternatives.

For the proof of Gärdenfors' theorem, the only weakening of weak-dominance manipulability that we need is given by the following.

Definition 8.1.4. A voting rule V is *almost weak-dominance manipulable* if it is weak-dominance manipulable or there exists a profile $\mathbf{P} = (R_1, \ldots, R_n)$, which we think of as giving the true preferences of the n voters, and in which voter i has x on top, y in second place, and z in third place, and another ballot S_i, which we think of as a disingenuous ballot from voter i such that, letting $\mathbf{P}' = (R_1, \ldots, R_{i-1}, S_i, R_{i+1}, \ldots, R_n)$,

$$V(\mathbf{P}) = \{x, y, z\} \quad \text{and} \quad V(\mathbf{P}') = \{x, y\}.$$

This brings us to the following:

Theorem 8.1.5. *In the context of non-linear ballots, suppose that A is a set of three or more alternatives,* $n \geq 3$*, and V is a voting rule for* (A, n) *that satisfies the CWC. Then V is almost weak-dominance manipulable.*

Gärdenfors' original theorem also assumed that the voting rule was both anonymous and neutral (meaning that it treats all alternatives the same and all voters the same). These assumptions, it turns out, are unnecessary. We will, at the appropriate place in the argument, indicate where Gärdenfors made use of these assumptions and how we modify the argument to render them unnecessary.

Proof: We need several claims.

Claim 1: Let $\mathbf{P}_1$ be the profile in which voters 1, 2, and 3 rank the three alternatives x, y, and z as in the voters' paradox:

x	y	z
y	z	x
z	x	y

with all other alternatives tied and below these three, and every other voter having all the alternatives tied on their ballots. Then $V(\mathbf{P}_1) \supseteq \{x, y, z\}$.

Proof. If the winner were x, together with any alternatives other than y and z, then voter 2 could interchange y and z and make z the Condorcet winner, thus making $\{z\}$ the winner of the election, an improvement. A similar argument shows that the winner can't be y or z, with or without alternatives other than x, y, and z.

If the winners were x and y, together with any alternatives other than x, y, and z, then voter 3 could interchange x and z on his or her ballot and make $\{x\}$ the Condorcet winner, thus making $\{x\}$ the winner of the election, an improvement. A similar argument shows that the winner can't be x and z or y and z, together with any alternatives other than x, y, and z.

Finally, if none of x, y, and z were winners, then voter 1, say, could gain by interchanging x and y on his or her ballot, thus making y a Condorcet winner. Hence, the winners must be x, y, and z, possibly together with some alternatives other than x, y, and z.

Claim 2. Let $\mathbf{P}_2$ be the profile in which voters 1, 2, and 3 rank the three alternatives x, y, and z as follows:

x	yz	z
y		x
z	x	y

with all other alternatives tied and below these three, and every other voter having all the alternatives tied on their ballots. Then $V(\mathbf{P}_2) = \{z\}$.

Proof. If $x \in V(\mathbf{P}_2)$, then voter 2 could move z down and make y the Condorcet winner, thus making $\{y\}$ the winner of the election, an improvement. Voter 2 could do the same thing if the set of winners contained at least one alternative other than x, y, and z, together with either y, z, or y and z. If $V(\mathbf{P}_2) = \{y\}$ or $V(\mathbf{P}_2) = \{y, z\}$, then voter 2 could gain by switching from his or her ballot in $\mathbf{P}_1$ to his ballot in $\mathbf{P}_2$. This leaves only $V(\mathbf{P}_2) = \{z\}$, as desired.

Claim 3. Let $\mathbf{P}_3$ be the profile in which voters 1, 2, and 3 rank the three alternatives x, y, and z as follows:

$$\begin{array}{ccc} y & yz & z \\ x & & x \\ z & x & y \end{array}$$

with all other alternatives tied and below these three, and every other voter having all the alternatives tied on their ballots. Then $V(\mathbf{P}_3) = \{z\}$.

Proof. If $x \in V(\mathbf{P}_3)$, then voter 2 could move z down and make y the Condorcet winner, thus making $\{y\}$ the winner of the election, an improvement. Again, voter 2 could do the same thing if the set of winners contained at least one alternative other than x, y, and z, together with either y, z, or y and z. If $V(\mathbf{P}_3) = \{y\}$ or $V(\mathbf{P}_3) = \{y, z\}$, then voter 1 could gain by switching from his or her ballot in $\mathbf{P}_2$ to his or her ballot in $\mathbf{P}_3$. Thus, $V(\mathbf{P}_3) = \{z\}$, as desired.

This is the point in the argument at which Gärdenfors invoked anonymity and neutrality to show that $V(\mathbf{P}_3) = \{y, z\}$, a contradiction. But this is not necessary, because we can simply repeat the above three claims on the profiles we get by interchanging y with z and interchanging voter 1's ballot with voter 3's ballots. That is, if we let $\mathbf{P}'_1$, $\mathbf{P}'_2$, and $\mathbf{P}'_3$ be the following:

$$\begin{array}{ccc} y & z & x \\ x & y & z \\ z & x & y \end{array} \qquad \begin{array}{ccc} y & yz & x \\ x & & z \\ z & x & y \end{array} \qquad \begin{array}{ccc} y & yz & z \\ x & & x \\ z & x & y \end{array}$$

then we can repeat the arguments from above to conclude that $V(\mathbf{P}'_3) = \{y\}$. But $\mathbf{P}_3 = \mathbf{P}'_3$, and this is the desired contradiction. □

We conclude this section by taking slight exception, as we said we would, to Gärdenfors' characterization of almost-weak-dominance manipulation as a "sure-thing principle." Consider the following ballots:

$$\begin{array}{ccccc} a & a & b & c & c \\ b & c & a & b & b \\ c & b & c & a & a \end{array}$$

Suppose that ties are to be broken by sequential pairwise voting (as we did in Section 4.4) with the alternatives pitted against each other in the order $c\,b\,a$.

If there are other alternative that can be moved around, voter 1 would like to manipulate the situation so that the winning set is $\{a, b, c\}$ instead of $\{a, b\}$ – exactly the opposite of what Gärdenfors' "sure-thing principle" would dictate. The point is that with a runoff using one-on-one contests with the order cba, we have a emerging as the winner from the set $\{a, b, c\}$, but b emerging as the winner from the set $\{a, b\}$.

8.2 Characterization Theorems for Non-Resolute Voting Rules

The version of the Gibbard–Satterthwaite theorem that we presented characterized non-manipulable voting rules that are resolute in the context of linear ballots. In this section, we derive a version of this result for non-resolute voting rules that uses the Duggan–Schwartz theorem and that generalizes the resolute case. Our starting point is the following definition from Duggan and Schwartz (2000).

Definition 8.2.1. In the context of linear ballots, a voting rule V satisfies *residual resoluteness* (*RR*) provided that $|\mathrm{V}(\mathbf{P})| = 1$ whenever **P** is a profile in which there are two alternatives x and y such that $\{x, y\}$ is a top set and all but at most one voter has y over x.

The condition RR allows us to replace "nominator" by "dictator" in the conclusion of the Duggan–Schwartz theorem.

Theorem 8.2.2 (Duggan and Schwartz, 2000). *In the context of linear ballots, if* n *is a positive integer and* A *is set of three or more alternatives, then any voting rule for* (A, n) *that is non-imposed,*[26] *satisfies RR, and cannot be manipulated by an optimist or a pessimist is a dictatorship.*

Proof: We first claim that if $\mathbf{P}'$ is a profile in which every voter has x on top and y second, then $\mathrm{V}(\mathbf{P}') = \{x\}$. To see this, note first that if $\mathrm{V}(\mathbf{P}')$ did not contain x, we could choose **P** such that $x \in \mathrm{V}(\mathbf{P})$ and then change $\mathbf{P}'$ to **P** one ballot at a time until x appeared as a winner, thus allowing some voter to improve his or her max to his most preferred alternative x. But V satisfies RR, so $|\mathrm{V}(\mathbf{P}')| = 1$. Thus, because $x \in \mathrm{V}(\mathbf{P}')$, we can conclude that $\mathrm{V}(\mathbf{P}') = \{x\}$.

Theorem 4.1.2 now guarantees that there is a voter i whose top choice is among the winners. Suppose, for contradiction, that there is a profile **P** such

[26] The proof will show that the assumption that V is non-imposed can be weakened to just assuming every alternative x is among the winners (but not necessarily a singleton winner) for at least one profile.

that voter i has x on top of his or her ballot, but $V(\mathbf{P}) \neq \{x\}$. Fixing the ballots of the other voters, choose a ballot for voter i such that an alternative y occurs in $V(\mathbf{P})$ that is as low on his or her ballot as possible. Let $\mathbf{P}'$ be any profile in which voter i has x on top of his or her ballot and y second, and every other voter has y on top and x second. By our assumption, $|V(\mathbf{P}')| = 1$, and, because x is on top of voter i's ballot, $x \in V(\mathbf{P}')$. Thus, $V(\mathbf{P}') = \{x\}$.

Now change $\mathbf{P}'$ to $\mathbf{P}$ one ballot at a time for every voter except voter i. If y appears at some point, then that voter has improved his or her max to the most preferred alternative y. If y never appears, let $\mathbf{P}''$ be the resulting profile, and note that $\mathbf{P}''$ and $\mathbf{P}$ differ only because of voter i's ballot. But now, if voter i's ballot in $\mathbf{P}$ represents his or her true preferences, then he or she can use the disingenuous ballot in $\mathbf{P}''$ to improve his or her min from y to something better (it is better because we chose y to be as low as possible). □

We can now characterize the voting rules that satisfy RR and are non-manipulable.

Theorem 8.2.3. *In the context of linear ballots, if* n *is a positive integer and A is a set of three or more alternatives, then a voting rule V for* (A, n) *is non-manipulable by optimists and pessimists and satisfies RR iff one of the following holds:*

(1) *There are two alternatives* x *and* y *and two simple games* $G_x = (N, E_x)$ *and* $G_y = (N, G_y)$ *that are "pairwise proper" in the sense that if* $X \in E_x$ *and* $Y \in E_y$*, then* $X \cap Y \neq \emptyset$*, and for which every singleton set is an edge in one of the games or its complement is an edge in the other, and such that* {x} *wins if the set of voters who rank* x *over* y *is an edge in* G_x, {y} *wins if the set of voters who rank* y *over* x *is an edge in* G_y*, and* {x, y} *wins otherwise. Notice that if* $G_x = (N, \wp(N))$ *and* $G_y = (N, \emptyset)$*, then* x *is the winner regardless of the ballots.*
(2) *There is a set B containing three or more alternatives, and a particular voter such that the unique winner of the election is the element of B that is ranked highest by this voter.*

Proof: It is easy to see that the voting rules described in the theorem are all nonmanipulable by optimists and pessimists and satisfy RR. For the converse, let B be the set of "viable" alternatives in the sense that there is at least one profile that yields that alternative as one of the winners. If B is a singleton, then the system is as described in (1) of the theorem.

If B has exactly two elements then it is easy to see that (2) in the theorem holds (but this requires using non-manipulability to show that the placement of other alternatives on the ballots has no effect on whether a wins or b wins).

Suppose now that B has at least three alternatives. We first claim that if $x \in$ B, then V(**P**) $= \{x\}$ whenever **P** is a profile in which every voter ranks x first and some other common element y second. This is because RR guarantees that $|$V(**P**)$| = 1$ and, if the result were $\{y\}$, we could convert ballots one by one until the outcome included x, thus improving the max for that voter.

Let V′ be the voting system on the set B obtained by applying the original voting system to the result of adding all the alternatives not in B to the bottom (in some fixed predetermined order) of all the ballots. Then V′ is still non-manipulable, RR still holds, and by the argument in the previous paragraph, for every x in B we have V′(**P**) $= \{x\}$ for at least one profile **P**, and so V′ is non-imposed.

It now follows from the Duggan–Schwartz theorem that there is a voter i who is a nominator for V′ in the sense that the top-ranked alternative on his or her ballot is among the winners. We claim that the winner in the original system is a singleton set consisting of the element of B ranked highest by the weak dictator for V′. Suppose not, and choose a profile **P** such that the set V(**P**) of winners includes some alternative x, necessarily in B, that is ranked lower on the weak dictator's ballot than some other element y of B.

First of all, move all the alternatives not in B below x on the weak dictator's ballot. If the winner switches to $\{y\}$, then the dictator has improved his or her min. Otherwise, the new min is some x' that is no higher on the dictator's list than x. The result is that y is on top of the dictator's list. Now we can, one ballot at a time, move x to the top of each of the other ballots. Then x remains a winner, or undoing the last change before x became a loser would improve the max for that voter. Similarly, if we now move y into second place on each of these ballots (other than the dictator's), then x stays a winner. Finally, move x into the second spot on the dictator's list. By RR, the winning set is now a singleton. If V(**P**) $= \{y\}$, then the dictator has improved his or her min, and that's impossible. So V(**P**) $= \{x\}$, and we can then use down-monotonicity for singleton winners to place all alternatives not in B at the bottom of all the ballots in the correct order. But now V(**P**) $= \{y\}$, and this is a contradiction. □

8.3 Another Feldman Theorem

Our focus in this section is the extent to which the assumptions in the theorems of Barberá and Kelly can be weakened and yet still yield the existence of what might be called a "local nominator" in the sense of a voter being able to ensure that x is one of the winners from a *particular* two-element agenda $\{x, y\}$, but not necessarily for *every* two-element agenda. Of course, if one also assumes that all alternatives are treated the same (neutrality), then any local nominator is a pairwise nominator.

The result we present is a fairly easy observation of Feldman (1979a), based on some work of Blau and Deb (1977). It shows that if V is downward-normal and non-manipulable on two-element agendas, then there exists a local nominator for V, provided that – and this is a big assumption – there are at least as many alternatives as there are voters.

Throughout this section we assume that V is a social choice function and that we are in the context of linear ballots. Our starting point is to compare Theorem 5.1.6 and Corollary 5.3.4, both of which impose additional assumptions on V that allow us to conclude (in particular) that there is a pairwise nominator for V.

The difference in the two results is that, although both assume that V is pairwise non-imposed and non-manipulable for two-element agendas, Theorem 5.1.6 assumes additionally that V is quasitransitive, and Corollary 5.3.4 assumes that V is normal and non-manipulable for three-element agendas as well as two-element agendas. The following example shows there are limitations to how these two results can be strengthened.

Example 8.3.1. Assume $|A| = 3$ and $|N| = 4$. For any profile $\mathbf{P}$ and alternatives $x, y \in A$, set

$$V(\mathbf{P})(\{x, y\}) = \begin{cases} \{x\} & \text{if at least 3 of the 4 voters rank } x \text{ over } y \\ \{y\} & \text{if at least 3 of the 4 voters rank } y \text{ over } x \\ \{x, y\} & \text{otherwise.} \end{cases}$$

For $|v| = 3$, set $V(\mathbf{P})(v) = \{x \in v \colon \forall y \in v, x \in V(\mathbf{P})(\{x, y\})\}$.

Notice that the social choice function given in Example 8.3.1 is, in fact, well-defined in the sense that if $\mathbf{P}$ is a profile and v is a three-element agenda (there is actually only one three-element agenda), then $V(\mathbf{P})(v) \neq \emptyset$. That is, if $x \notin V(\mathbf{P})(v)$, then there is an alternative y such that at least three of the four voters rank y over x. Now, if $y \notin V(\mathbf{P})(v)$, then at least three of the four voters rank the remaining alternative z over y. But then we definitely have $z \in V(\mathbf{P})(v)$, because at least two of the four voters rank z over x.

It is easy to see that the V of Example 8.3.1 is pairwise non-imposed and non-manipulable for two-element agendas. Moreover, V is normal, but there is no local nominator for V.

It now follows from Corollary 5.3.4 that V is not quasitransitive, as can be seen directly from the following profile $\mathbf{P}$ in which the base relation has x strictly preferred to y and y strictly preferred to z, but x not strictly preferred to z.

z	x	x	y
x	y	y	z
y	z	z	x

Similarly, Theorem 5.1.6 lets us conclude that V must be manipulable for three-element agendas, something we can again directly verify using the profile **P** above. That is $V(\mathbf{P})(\{x, y, z\}) = \{x\}$, but if the last voter interchanges y and z, then we get the new profile $\mathbf{P}'$:

z	x	x	z
x	y	y	y
y	z	z	x

Now $V(\mathbf{P}')(\{x, y, z\}) = \{x, z\}$, and with true preferences y over z over x, this voter prefers $\{x, z\}$ to $\{x\}$.

Notice, however, that in Example 8.3.1, $|A| < |N|$. This brings us to the main result of this section, which is, as we said, an observation of Feldman based on some work of Blau and Deb.

Theorem 8.3.2. *If a social choice function V is non-manipulable for two-element agendas and downward normal, and if $|A| \geq |N|$, then there exists a local nominator for V.*

Proof: We first observe that if V fails to be pairwise-non-imposed, then there exists an ordered pair (x, y) of distinct alternatives such that for every profile $\mathbf{P}$, $x \in V(\mathbf{P})(\{x, y\})$. But this means that every voter can use x to block y, and so we are done in this case.

Now because $|A| \geq |N|$, we can choose distinct alternatives $x_1, \ldots, x_n$ (where $|N| = n$) and consider the following profile **P**:

x_1	x_2	x_3	.	.	.	x_n
x_2	x_3	x_4	.	.	.	x_1
x_3	x_4	x_5	.	.	.	x_2
.	.	.	.	.	.	.
.	.	.	.	.	.	.
.	.	.	.	.	.	.
x_{n-1}	x_n	x_1	.	.	.	x_{n-2}
x_n	x_1	x_2	.	.	.	x_{n-1}
.	.	.	.	.	.	.
.	.	.	.	.	.	.
.	.	.	.	.	.	.

Because $\mathrm{V}(\mathbf{P})(\{x_1, \ldots, x_n\}) \neq \emptyset$, we can choose i such that $x_i \in \mathrm{V}(\mathbf{P})(\{x_1, \ldots, x_n\})$. Now let $x = x_i$ and let $y = x_{i-1}$ (with the understanding that if $i = 1$, then $i - 1 = n$). With these choices, voter i is the only voter to rank x over y, with every other voter ranking y over x, and yet $x \in \mathrm{V}(\mathbf{P})(\{x_1, \ldots, x_n\})$. Because V is downward normal, we have that $x \in \mathrm{V}(\mathbf{P})(\{x, y\})$.

To complete the proof, we need only establish that if $\mathbf{P}'$ is any other profile in which voter i ranks x over y and everyone else ranks x and y arbitrarily, then $x \in \mathrm{V}(\mathbf{P}')(\{x, y\})$. But this follows immediately from two observations based on V's being non-manipulable on two-element agendas and pairwise non-imposed:

(1) V satisfies binary IIA, as demonstrated in the proof of Theorem 5.3.3
(2) V satisfies monotonicity, as demonstrated in the proof of case 3 of Lemma 5.1.9

This completes the proof of Theorem 8.3.2. □

8.4 Characterization Theorems for Non-Resolute Social Choice Functions

In this section we assume for simplicity that ballots are linear, and we investigate the effect of imposing some fairly strong democratic assumptions on a social choice function V. The starting point is the observation that if V is non-manipulable on two-element agendas, then V satisfies binary IIA. This means that for each ordered pair (x, y) of distinct alternatives, there is a collection $\mathrm{E}(x, y)$ of subsets of N such that, for any profile $\mathbf{P}$,

$$\mathrm{V}(\mathbf{P})(\{\mathrm{x}, \mathrm{y}\}) = \begin{cases} \{x\} & \text{if } \{i \in N;\ \text{voter } i \text{ ranks } x \text{ over } y\} \in \mathrm{E}(x, y) \\ \{y\} & \text{if } \{i \in N;\ \text{voter } i \text{ ranks } y \text{ over } x\} \in \mathrm{E}(y, x) \\ \{x, y\} & \text{otherwise.} \end{cases}$$

Moreover, any collection of sets $\{\mathrm{E}(x, y) : x \neq y\}$ yields a social choice function defined on $[\mathrm{V}]^2$, subject only to the requirement that one never have $\mathrm{X} \in \mathrm{E}(x, y)$ and $\mathrm{N} - \mathrm{X} \in \mathrm{E}(y, x)$.

Let's now see what effect various assumptions have on the collections $\mathrm{E}(x, y)$, beginning with neutrality. If there were literally only two alternatives, neutrality would require that $\mathrm{E}(x, y) = \mathrm{E}(y, x)$. But with more than two alternatives the idea of treating all alternatives the same is better captured by the notion from Feldman (1979a), and this notion is equivalent to requiring that there exist a single collection E such that $\mathrm{E}(x, y) = \mathrm{E}$ for every ordered pair (x, y) of distinct alternatives. Thus, neutrality means that, on two-element agendas, V is determined by a hypergraph H = (N, E) that is proper in the sense that one never has both X and N − X in E.

Setting neutrality aside for the moment, let's assume only that V is non-manipulable on two-element agendas. This requires that V must be monotone, and this happens iff the collections $E(x, y)$ are closed under the formation of supersets. Hence, on two-element agendas, V is given by a collection of not just hypergraphs, but proper simple games.

Finally, if V is anonymous, then, without assuming either neutrality or non-manipulability on two-element agendas, it is easy to see that whether or not a set X belongs to some $E(x, y)$ must depend only on the cardinality of X. This means that for each pair (x, y) there is a number $q = q(x, y)$ such that $n/2 < q \leq n + 1$ and $E(x, y) = \{X \subseteq N \colon |X| \geq q\}$. Notice that if V is pairwise non-imposed, $q = n + 1$ is ruled out.

Putting together what we have so far yields the following:

Theorem 8.4.1. *In the context of linear ballots, if A is a set of three or more alternatives and* n *is a positive integer, then the following are equivalent for any social choice function V for* (A, n)*:*

(1) V is neutral, anonymous, and non-manipulable on two-element agendas.
(2) There is a number q *with* $n/2 < q \leq n + 1$ *such that, for every profile* $\boldsymbol{P}$*, and every pair* x, y *of distinct alternatives,*

$$V(\boldsymbol{P})(\{x, y\}) = \begin{cases} \{x\} & \text{if at least } q \text{ voters rank } x \text{ over } y \\ \{y\} & \text{if at least } q \text{ voters rank } y \text{ over } x \\ \{x, y\} & \text{otherwise.} \end{cases}$$

Notice that Example 8.3.1 is the special case of (2) in which $n = 4$ and $q = 3$. In particular this means that with no additional assumptions beyond those in part (1) of Theorem 8.4.1, nothing can be said about nominators. Moreover, the only way there can be nominators in the procedure from (2) of Theorem 8.4.1 is if $q = n$.

Thus, if $|A| \geq |N|$, then Corollary 5.3.4 can be strengthened by replacing non-manipulable by non-manipulable on two-element agendas, if at the same time, we require anonymity and neutrality. This strengthens a theorem in Feldman (1979a).

Theorem 8.4.2. *In the context of linear ballots, if A is a set of three or more alternatives and* $n \leq |A|$*, then the following are equivalent for any social choice* *function V for* (A, n)*:*

(1) V is neutral, anonymous, normal, and non-manipulable on two element agendas.

(2) V is the Pareto rule: For every profile **P** *and every agenda* v,

$$x \in V(\boldsymbol{P})(v) \textit{ iff } x \in v \textit{ and } \forall y \in v\, \exists i \textit{ such that } x P_i y.$$

9

Other Election-Theoretic Contexts

9.1 Introduction

Throughout the book, we have uniformly imposed certain restrictions on the election-theoretic contexts that we considered. In particular, we assumed the following:

(1) Ballots are lists (with or without ties) of the set A of alternatives.
(2) An election outcome is either an alternative, a set of alternatives, a list, or a choice function.

In this concluding chapter, we briefly consider a few results that apply to situations that go beyond these assumptions.

In Sections 9.2 and 9.3, we consider two natural contexts in which a ballot is a set of alternatives. Here, we discuss an important voting system known as *approval voting*, and we present a very pretty result due to Barberà, Sonnenschein, and Zhou.

Finally, in Section 9.4, we present an elegant theorem of Allan Gibbard that takes place in the context wherein the outcome of an election is a "probability vector" that, in a very rough sense, gives the probability that each alternative will eventually emerge as the unique winner. Gibbard's theorem completely characterizes systems of this type that are non-manipulable.

9.2 Ballots That Are Sets: Approval Voting and Quota Systems

A natural election-theoretic context is the one in which a ballot is a set of alternatives, and the outcome of an election is also a set of alternatives. For the sake of mathematical naturality, we allow ballots and outcomes to be the empty set.

In this context, there are two natural kinds of voting procedures that are neutral, anonymous, and monotone in an obvious sense, but their description requires a bit of notation. We let $\mathbf{P} = (P_1, \ldots, P_n)$ denote a profile as before, but now P_i denotes a subset of the set A of alternatives. If $a \in A$, we let

$$NP(a) = |\{i \in N: a \in P_i\}|.$$

Thus, NP(a) is the number of voters who include the alternative a on their ballot in **P**.

Definition 9.2.1.

(1) *Approval voting* is the system V_A in which

$$V_A(\mathbf{P}) = \{a \in A: \forall b \in A, NP(a) \geq NP(b)\}.$$

(2) For a positive integer q, the *quota system* V_q is the system in which

$$V_q(\mathbf{P}) = \{a \in A: NP(a) \geq q\}.$$

Approval voting arose in the 1970s, and although proposed independently by a number of people, its foremost advocates and investigators were Steven J. Brams, a political scientist at New York University, and Peter C. Fishburn, a mathematician at Bell Labs in Murray Hill, New Jersey. Their 1982 text provides an excellent exposition, but we offer a few comments here.

Approval voting is used to elect the secretary-general of the United Nations and by several academic and professional societies including the 400,000-member Institute of Electrical and Electronics Engineers (IEEE) and the National Academy of Sciences. Part of the appeal of approval voting is that it eliminates the question of whether one should worry about "throwing away" one's vote on a most-preferred candidate who has little chance of winning. For example, in the US presidential election of 2000, one could have voted for the set {Gore, Nader} if approval voting had been used. The intuition behind approval voting is that one is simply indicating the set of candidates that he or she finds "acceptable" for the position. Interestingly, when approval voting was used in Eastern Europe and the former Soviet Union, it was cast as "disapproval voting," with the fewest votes determining the winner.

Quota systems, if viewed appropriately, are also quite common. If a group is settling a number of (independent) yes–no questions by majority rule, then we can think of each voter as submitting, in a single ballot, the set of issues that he or she supports, with a quota equal to half the number of voters. Some academic departments use a quota system in hiring to reduce the pool of candidates from a relatively large number to a relatively small number (agreeing, for example, to further consider only those candidates approved of by one-third of the department).

Certainly, one of the advantages of approval voting over quota systems is that, with approval voting, the set of winners is typically much smaller than with a quota system. In fact, with a large number of voters and a small number of candidates, approval voting almost always yields a unique winner.

But when it comes to non-manipulability of the kind we are considering, there is a very easy result asserting that quota systems are strictly superior to approval voting.

Theorem 9.2.2. *Suppose we have a notion of when a voter prefers one set of alternatives to another and suppose that, with this notion, approval voting is non-manipulable. Then, with this same notion, every quota system is also non-manipulable.*

Proof: Suppose that voter i can manipulate a quota system V_q by submitting a disingenuous ballot $D \subseteq A$ to obtain an election outcome $Y \subseteq A$ that he or she prefers (according to our unspecified notion) to the election outcome $X \subseteq A$ that he or she obtains by submitting an honest ballot $H \subseteq A$. In passing from H to D, voter i might have deleted some alternatives in H that received more than q votes or fewer than q votes, but such deletions have no effect on the set of winners. Thus we lose no generality in assuming that the alternatives deleted in passing from H to D received a total of exactly q votes when voter i used H. Similarly, we lose no generality in assuming that the alternatives added in passing from H to D received exactly $q - 1$ votes when voter i used H.

But now, starting with the profile in which voter i used H, we can alter this profile by eliminating alternatives from other ballots until each alternative in X has exactly q votes. But now it is clear that if we were using approval voting on this new profile, voter i would still obtain X as the set of winners using the honest strategy H, and he or she would still obtain Y after changing to the strategy D. Hence, approval voting is subject to the same instance of manipulation as was the quota system. □

Quota systems, then, are at least as non-manipulable as approval voting. But is there a reasonable notion of when a voter prefers one set of alternatives to another that yields an interesting non-manipulability result for quota systems (and, at the same time, yields a situation in which approval voting can be manipulated but quota systems cannot)? The answer is yes, and it is adapted from the work of Barberá, Sonnenschein, and Zhou (1991).

9.3 The Barberá–Sonnenschein–Zhou Theorem

Ballots will be continue to be subsets of A, as will outcomes of elections. We will not, however, restrict attention to voting systems that are anonymous or neutral. In terms of preferences over subsets, we assume that each voter has a set A_i of alternatives that he or she "approves of" and a linear ordering $<_i$ of subsets of A that satisfies *separability*:

$$\emptyset <_i \{x\} \quad \text{iff} \quad x \in A_i \quad \text{iff} \quad X <_i X \cup \{x\}.$$

This notion of preference carries with it a more-is-better sense when dealing with sets of alternatives, all of which are approved of by a single voter. Thus, if a voter approves of both a and b, then he or she will prefer an election outcome of $\{a, b\}$ to $\{a\}$. There are several contexts in which this makes sense. For example, if each of several groups is to put forth several candidates, with a single winner to be eventually drawn by a lottery in which all candidates have the same chance of winning, then adding two candidates one approves of is strictly better than adding just one.

With this, we have the following:

Definition 9.3.1. A voting system V is *BSZ-manipulable* (Barberá–Sonnenschein–Zhou) iff there exists a profile $\mathbf{P} = (P_1, \ldots, P_{i-1}, A_i, P_{i+1}, \ldots, P_n)$ in which voter i's ballot is his or her set A_i and another ballot Q_i, which we think of as a disingenuous ballot from voter i such that, letting $\mathbf{P}' = (P_1, \ldots, P_{i-1}, Q_i, P_{i+1}, \ldots, P_n)$, we have

$$V(\mathbf{P}) <_i V(\mathbf{P}').$$

Our first observation is that approval voting is BSZ-manipulable, as can be seen by the following trivial example with two voters and two alternatives. If voter 1 approves of both a and b, and voter 2 casts $\{a\}$ as his or her ballot, then voter 1 gets $X = \{a\}$ with a sincere ballot $A_1 = \{a, b\}$, but gets $Y = \{a, b\}$ with a disingenuous ballot $Q_1 = \{b\}$.

The following is essentially Lemma 1 on page 601 of Barberá, Sonnenschein, and Zhou (1991); the full Barberá–Sonnenschein–Zhou theorem is considerably deeper.

Theorem 9.3.2. *For a voting system V, the following are equivalent:*

(1) V is BSZ non-manipulable.

(2) For each alternative a *there is a simple game $G_a = (N, W_a)$ such that, for every profile* $\boldsymbol{P}$, $a \in V(\boldsymbol{P})$ *iff* $\{i : a \in P_i\} \in W_a$.

Proof: The proof that (2) implies (1) is trivial. To prove that (1) implies (2), fix $a \in \mathrm{A}$ and let W_a be the collection of subsets of N wherein $\mathrm{X} \in \mathrm{W}_a$ iff there exists a profile **P** such that $a \in \mathrm{V}(\mathbf{P})$ and $\mathrm{X} = \{i \in \mathrm{N}: a \in \mathrm{P}_i\}$. We now claim that for any profile $\mathbf{P}'$, $a \in \mathrm{V}(\mathbf{P}')$ iff $\{i \in \mathrm{N}: a \in \mathrm{P}_i'\} \in \mathrm{W}_a$.

One direction is trivial: If $a \in \mathrm{V}(\mathbf{P}')$, then $\{i \in \mathrm{N}: a \in \mathrm{P}_i'\} \in \mathrm{W}_a$. For the converse, assume $\{i \in \mathrm{N}: a \in \mathrm{P}_i'\} \in \mathrm{W}_a$ and choose a profile **P** such that $\{i \in \mathrm{N}: a \in \mathrm{P}_i'\} = \{i \in \mathrm{N}: a \in \mathrm{P}_i\}$ and $a \in \mathrm{V}(\mathbf{P})$. We want to show that $a \in \mathrm{V}(\mathbf{P}')$. Suppose not. Then we can one by one change the ballots in **P** to those in $\mathbf{P}'$ until the change of a single ballot (say voter i) causes the election outcome to go from including a to excluding a, but voter i did not change his or her ballot regarding the inclusion or exclusion of a.

Suppose first that a is included in both of voter i's ballots. Take P_i' to represent an honest ballot for voter i and consider a situation in which voter i's ordering $<_i$ on subsets of A is determined by assigning weights to elements of A and then ordering these sets by total weight. If the weight of a is $2n$, and $0 <$ weight of $x < 1$ for every $x \in \mathrm{A}_i$, and $-1 <$ weight of $x < 0$ for every $x \in \mathrm{A} - \mathrm{A}_i$, then every set containing a is preferred to every set not containing a, and the resulting ordering is separable with respect to P_i'. But then this is an instance of manipulation.

On the other hand, if a is excluded from both of voter i's ballots, then we can take P_i to be voter i's sincere ballot, and assign weights as before, but with the weight of alternative a set at $-2n$. Again, this is an instance of manipulation. □

Notice that if V is given by a sequence (G_a: $a \in \mathrm{A}$) of simple games, then V is neutral iff there is a simple game G such that $\mathrm{G} = \mathrm{G}_a$ for every $a \in \mathrm{A}$. And V is anonymous iff for each a there is a number q_a such that $\mathrm{X} \in \mathrm{W}_a$ iff $|\mathrm{X}| \geq q_a$. In particular, V is neutral and anonymous iff V is a quota system. Thus, we have the following:

Corollary 9.3.3. *V is a quota system iff V is neutral, anonymous, and BSZ-non-manipulable.*

9.4 Outcomes That Are Probabilstic Vectors: Gibbard's Theorem

Allan Gibbard (1977) considered the voting-theoretic context in which a voting system V, given a linear profile **P**, produces an election-theoretic outcome that is a function $\mathrm{V}(\mathbf{P})\colon \mathrm{A} \to [0, 1]$ such that $\sum\{\mathrm{V}(\mathbf{P})(a)\colon a \in \mathrm{A}\} = 1$. The intuition here is that a single alternative will eventually be chosen via a random device designed so that an alternative $a \in \mathrm{A}$ has a probability of winning given by $\mathrm{V}(\mathbf{P})(a)$. Any such voting system will be called a *probabilistic voting system.*

There are two kinds of probabilistic voting systems that come into play in Gibbard's theorem. They are given by the following.

Definition 9.4.1. A probabilistic voting system V is said to be *unilateral* iff there is a voter i such that, for every pair of profiles $\mathbf{P}$ and $\mathbf{P}'$, $\mathrm{V}(\mathbf{P}) = \mathrm{V}(\mathbf{P}')$ whenever $\mathrm{P}_i = \mathrm{P}'_i$. V is said to be *duple* if there are alternatives x and y such that, for every profile $\mathbf{P}$, $\mathrm{V}(\mathbf{P})(a) = 0$ if $a \neq x, y$.

Probabilistic voting systems can be combined to yield a new system as follows:

Definition 9.4.2. The probabilistic voting system V is a *probabilisitic mixture* of the probabilistic voting systems $\mathrm{V}_1, \ldots, \mathrm{V}_m$ iff there exist $\alpha_1, \ldots, \alpha_m \in (0, 1]$ with $\sum \alpha_i = 1$, and, for every profile $\mathbf{P}$ and every $a \in \mathrm{A}$,

$$\mathrm{V}(\mathbf{P})(a) = \alpha_1 \mathrm{V}_1(\mathbf{P})(a) + \cdots + \alpha_m \mathrm{V}_m(\mathbf{P})(a).$$

The notion of manipulability used by Gibbard is the following. Recall that if P is a linear ballot, then a utility function u: $\mathrm{A} \to \Re$ *represents* P iff, for every $x, y \in \mathrm{A}$, $x\mathrm{P}y$ iff $u(x) > u(y)$.

Definition 9.4.3. A probabilistic voting system V is *expected-utility manipulable* if there exists a profile $\mathbf{P} = (\mathrm{P}_1, \ldots, \mathrm{P}_n)$, which we think of as giving the true preferences of the n voters, another ballot Q_i, which we think of as a disingenuous ballot from voter i, and a utility function u representing P_i (voter i's true preferences) such that, letting $\mathrm{V}(\mathbf{P}) = \mathrm{X}$ and $\mathrm{V}(\mathrm{P}_1, \ldots, \mathrm{P}_{i-1}, \mathrm{Q}_i, \mathrm{P}_{i+1}, \ldots, \mathrm{P}_n) = \mathrm{Y}$,

$$\sum\{\mathrm{V}(\mathbf{P})(x) \cdot u(x)\colon x \in \mathrm{X}\} < \sum\{\mathrm{V}(\mathbf{P})(y) \cdot u(y) : y \in \mathrm{Y}\}.$$

Thus, voter i's disingenuous ballot increases his or her expected utility according to the utility function u. There are two additional conditions we need before stating Gibbard's theorem. Recall that if P_i is a linear ballot and $\mathrm{X} \subseteq \mathrm{A}$, then X is a *top set for P_i* iff $\forall x \in \mathrm{X}\ \forall y \in \mathrm{A} - \mathrm{X}$, $x\mathrm{P}_{iy}$.

Definition 9.4.4. A probabilistic voting system V is *localized* if, for every profile $\mathbf{P}$ and every voter i, if X is top set for P_i, and if Q_i is another ballot for voter i in which X is also a top set (and thus Q_i is obtained from P_i by permuting elements in X and by permuting elements in A − X), then, letting $\mathbf{P}' = (\mathrm{P}_1, \ldots, \mathrm{P}_{i-1}, \mathrm{Q}_i, \mathrm{P}_{i+1}, \ldots, \mathrm{P}_n)$,

$$\sum\{\mathrm{V}(\mathbf{P})(x)\colon x \in \mathrm{X}\} = \sum\{\mathrm{V}(\mathbf{P}')(x)\colon x \in \mathrm{X}\}.$$

V is *monotone* if a voter's switching of an alternative with the alternative directly above it on his or her ballot does not decrease its probability.

With these notions in place, we can state Gibbard's theorem.

Theorem 9.4.5. *For a probabilistic voting system V, the following are equivalent:*

(1) V is non-manipulable.
(2) V is localized and monotonic.
(3) V is a probabilistic mixture of localized, monotonic probabilistic voting systems, each of which is either unilateral or duple.

For a proof of Theorem 9.4.5, see Gibbard (1977). Other treatments can be found in Duggan (1996), Nandeibam (1998), and Tanaka (2004).

References

Armstrong, T. *Arrow's theorem with restricted coalition algebras. Journal of Mathematical Economics* **7** (1980), 55–75.

Arrow, K. *A difficulty in the concept of social welfare. The Journal of Political Economy* **58** (1950), 328–46.

Arrow, K. *Social choice and individual values* (2nd ed.). Yale University Press: New Haven, 1963.

Arrow, K. and A. Sen and K. Suzumura (Eds.), *Handbook of social choice and welfare, vol. I*, North-Holland, New York, 2002.

Austen-Smith, D. and J. Banks, *Positive political theory I: collective preferences.* University of Michigan Press: Ann Arbor, 2000.

Barberá, S. *Manipulation of social choice mechanisms that do not leave 'too much' to chance. Econometrica* **45** (1977a), 1573–88.

Barberá, S. *Manipulation of social decision functions. Journal of Economic Theory* **15** (1977b), 266–78.

Barberá, S. *Pivotal voters. A new proof of Arrow's theorem. Economic Letters* **6** (1980), 13–16.

Barberá, S. *Strategy-proofness and pivotal voters: a direct proof of the Gibbard-Satterthwaite theorem. International Economic Review* **24** (1983), 413–17.

Barberá, S. and B. Dutta and Arunava Sen, *Strategy-proof set valued social choice functions. Journal of Economic Theory* **101** (2001), 374–94.

Barberá, S., H. Sonnenschein, and L. Zhou, *Voting by committees. Econometrica* **59** (1991), 595–609.

Bartholdi, J. and J. Orlin, *Single-transferable vote resists strategic voting. Social Choice and Welfare* **8** (1991), 341–54.

Batteau, P, J. Blin, and B. Monjardet, *Stability of aggregation procedures, ultrafilters, and simple games. Econometrica* **49** (1981), 527–34.

Beja, A. *Arrow and Gibbard-Satterthwaite theorem re-visited. Extended domains and shorter proofs. Mathematical Social Sciences* **25** (1993), 281–6.

Bell, J. and A. Slomson, *Models and ultraproducts: An Introduction.* North Holland: Amsterdam-London, 1969.

Benoit, J. *Strategic manipulation in voting games when lotteries and ties are permitted. Journal of Economic Theory* **102** (2002), 421–36.

Benoit, J. *The Gibbard-Satterthwaite theorem: a simple proof. Economic Letters* **69** (2000), 319–22.

Black, D. *Theory of committees and elections*. Cambridge University Press, Cambridge, 1958.

Blau, J. *Social choice functions and simple games. Bulletin of the American Mathematical Society* **63** (1957), 243–4.

Blau, J. *A direct proof of Arrow's theorem. Econometrica* **40** (1972), 61–7.

Blau, J. and R. Deb, *Social decision functions and the veto. Econometrica* **45** (1977), 471–82.

Brams, S. and P. Fishburn, *Approval voting*. Birkhäuser Boston, Cambridge, MA, 1983.

Brams, S. and P. Fishburn, *Voting procedures*. In the Handbook of Social Choice and Welfare. Arrow, Sen, and Suzumura, eds. (2002), 175–236.

Burani, N. and W. Zwicker, *Coalition formation games with separable preferences* (preprint). Department of Mathematics, Union College (2000).

Campbell, D. *Equity, efficiency, and social choice*. Clarendon Press, Oxford, 1992.

Campbell, D. and J. Kelly. *A trade-off result for preference revelation. Journal of Mathematical Economics* **34** (2000), 129–42.

Campbell, D. and J. Kelly. *A leximin characterization of strategy-proof non-resolute social choice procedures. Economic Theory* **20** (2002), 809–29.

Ching, S. and L. Zhou, *Multi-valued strategy-proof social choice rules. Social Choice and Welfare* **19** (2002), 569–80.

COMAP [Consortium for Mathematics and Its Applications] *For all practical purposes: Introduction to contemporary mathematics* (6th ed.) W. H. Freeman: New York, 2003.

Comfort, W. and S. Negrepontis, *The theory of ultrafilters*. Springer-Verlag: New York, 1974.

Condorcet, M. *Essai sur l'application de l' analyse 'a la probabiliti'e des d'ecisions rendues 'a la pluralit'e des voix*. De L'Imprimerie royale, Paris, (1785).

deFinetti, B. *La pr'evision ses lois logiques, ses sources subjectives. Ann. Inst. H. Poincaré* **7** (1937), 1–68.

Denicolo, V. *Independent social choice functions are dictatorial. Economics Letters* **19** (1985), 9–12.

Denicolo V. *Fixed agenda social choice theory: correspondence and impossibility theorems for social choice correspondences and social decision functions. Journal of Economic Theory* **59** (1993), 324–32.

Duggan, J. *A geometric proof of Gibbard's random dictator theorem. Economic Theory* **7** (1996), 365–9.

Duggan, J. and T. Schwartz, *Strategic manipulability is inescapable: Gibbard-Satterthwaite without resoluteness* (preprint). Department of Economics, University of Rochester, (1993).

Duggan, J. and T. Schwartz, *Strategic manipulability without resoluteness or shared beliefs: Gibbard-Satterthwaite generalized. Social Choice and Welfare* **17** (2000), 85–93.

Feldman, A. *Non-manipulable multi-valued social decision functions. Public Choice* **34** (1979a), 177–88.

Feldman, A. *Manipulation and the Pareto rule. Journal of Economic Theory* **21** (1979b), 473–82.

Feldman, A. *Manipulating voting procedures. Economic Inquiry* **17** (1979c), 452–72.

Feldman, A. *Strongly nonmanipulable multi-valued collective choice rules. Public Choice* **35** (1980a), 503–9.

Feldman, A. *Welfare economics and social choice theory*. Kluwer: Nijhoff, 1980b.

Felsenthal, D. and M. Machover, *The measurement of voting power: theory and practice, problems and paradoxes*. Edward Elgar: Cheltenham, UK, 1998.

Felsenthal, D. and M. Machover, *After two centuries, should Condorcet's voting procedure be implemented? Behavorial Sciences* **37** (1992), 250–74.

Fishburn, P. *Arrow's impossibility theorem: concise proof and infinite voters. Journal of Economic Theory* **2** (1970), 103–6.

Fishburn, P. *The theory of social choice*. Princeton University Press: Princeton, NJ, 1973.

Fishburn, P. *The axioms of subjective probability. Statist. Sci.* **1**(1986), 335–58.

Gärdenfors, P. *Manipulation of social choice functions. J. Econom. Theory* **13** (1976), 217–28.

Gärdenfors, P. *A concise proof of theorem on manipulation of social choice functions. Public Choice* **32** (1977), 137–42.

Gärdenfors, P. *On definitions of manipulation of social choice functions*. Aggregation and Revelation of Preferences, edited by Jean-Jacques Laffont. North Holland, Amsterdam, 1979.

Geanakoplos. *Three brief proofs of Arrow's impossibility theorem.* Cowles Discussion Paper 1123R, 1996.

Gibbard, A. *Manipulation of voting schemes: a general result. Econometrica* **41** (1973), 587–601.

Gibbard, A. *Manipulation of schemes that mix voting with chance. Econometrica* **45** (1977) 665–81.

Gibbard, A. *Straightforwardness of game forms with lotteries as outcomes. Econometrica* **46** (1978), 595–614.

Guilbaud, G. *Les théories de l'intérêt général et le problème logique de lagrégation. Economie Appliquée* **5** (1952), 501–84.

Hansson, B. *Group preferences. Econometrica* **37** (1969), 50–4.

Kelly, J. *Strategy-proofness and social choice functions without single-valuedness. Econometrica* **45** (1977), 439–46.

Kelly, J. *Arrow impossibility theorems*. Academic Press: New York, 1978.

Kelly, J. *Social choice theory: an introduction*. Springer-Verlag: New York, 1987.

Kelly, J. *Social choice bibliography. Social Choice and Welfare* **8** (1991), 97–169.

Kirman, A. and D. Sondermann, *Arrow's theorem, many agents and invisible dictators. Journal of Economic Theory* **5**, (1972), 267–77.

Lauwers, L. and L. Van Liedekerke. *Ultraproducts and aggregation. Journal of Mathematical Economics* **24** (1995), 217–37.

MacIntyre, I. and P. Pattanaik, *Strategic voting under minimally binary group decision functions. Journal of Economic Theory* **25** (1981), 338–52.

Makinson, D. *Combinatorial versus decision-theoretic components of impossibility theorems. Theory and Decision* **40** (1996), 181–90.

May, K. *A set of independent, necessary and sufficient conditions for simple majority decision. Econometrica* **20** (1952), 680–4.

McLean, I. and A. Urken (ed. and transl.), *Classics of social choice*. Michigan University Press: Ann Arbor, MI, 1993.

Merlin, V. and D. Saari, *A geometric examination of the Kemeny rule. Social Choice and Welfare* **17** (2000), 403–38.

Mihara, H. *Coalitionally strategyproof functions depend only on the most-preferred alternative. Social Choice and Welfare* **17** (2000), 393–402.

Mihara, H. *Existence of a coalitionally strategyproof social choice function: A constructive proof. Social Choice and Welfare* **18** (2001), 543–53.

Mill, J. *Considerations on representative government*. Harper and Brothers: New York, 1862.

Monjardet, B. *Une autre prevue du théorème d'Arrow. R.A.I.R.O.* **12** (1978), 291–6.

Monjardet, B. *Introduction to social choice theory: The Arrow and Gibbard-Satterthwaite theorem*. Cahiers MSE: CERMSEM. Université Paris 1, (1999).

Monjardet, B. *Social choice and the "Centre de Mathématique Sociale": Some historical notes. Social Choice and Welfare* (to appear in 2005).

Moulin, H. *The proportional veto principle. Rev. Econ. Stud.* **48** (1981), 407–16.

Moulin, H. *The strategy of social choice*. North-Holland: New York, 1983.

Moulin, H. *Fairness and strategy in voting. Proceedings of Symposia in Applied Mathematics* **33** (1985), 109–42.

Moulin, H. *Fair division and collective welfare*. The MIT Press: Cambridge, MA, 2003.

Muller, E. and M. Satterthwaite. *The equivalence of strong positive association and strategy proofness. Journal of Economic Theory* **14** (1977), 412–18.

Nandeibaum, S. *An alternative proof of Gibbard's random dictatorship result. Social Choice and Welfare* **15** (1998), 509–19.

Nurmi, H. *Comparing voting systems*. D. Reidel Publishing Company: Dordrecht, Holland, 1987.

Pazner, E. and E, Wesley, *Stability properties of social choices in infinitely large societies. Journal of Economic Theory* **14** (1977), 252–62.

Pazner, E. and E. Wesley, *Cheatproofness of the plurality rule in large societies. Review of Economic Studies* **45** (1978), 85–91.

Ramamurthy, K. *Coherent structures and simple games*. Kluwer: Dordrecht, Netherlands, 1990.

Riker, W. *Liberalism against populism: a confrontation between the theory of democracy and the theory of social choice*. W. H. Freeman: San Francisco, 1982.

Riker, W. *The art of political manipulation*. Yale University Press: New Haven and London, 1986.

Saari, D. *The geometry of voting*. Springer-Verlag: New York, 1994.

Saari, D. *Basic geometry of voting*. Springer-Verlag: New York, 1995.

Saari, D. *Choatic elections: a mathematician looks at voting*. The American Mathematical Society, 2001.

Satterthwaite, M. *Strategy-proofness and Arrow's conditions: existence and correspondence theorems for voting procedures and social welfare functions. Journal of Economic Theory* **10** (1975), 187–217.

Savage, L. *The foundations of statistics*. Wiley: New York, 1954.

Schmeidler, D. and H. Sonnenschein, *Two proofs of the Gibbard-Satterthwaite theorem on the possibility of a strategy-proof social choice function*, in *Decision Theory and Social Ethics Issues in Social Choice*. H. Gottinger and W. Leinfellner, editors. Reidel: Dordrecht (1978), 227–34.

Schofield, N. *Social choice and democracy*. Springer Verlag: Berlin, 1985.

Sen, A. *A possibility theorem on majority decisions. Econometrica* **34** (1966), 491–9.
Sen, A. *Collective choice and social welfare*. Holden Day: San Francisco, 1970.
Sen A. *Social choice theory: a re-examination. Econometrica* **45** (1977), 53–89.
Sen, A. *Choice, welfare, and measurement*. MIT Press: Cambridge, MA, 1982.
Sen, Arunava, *Another direct proof of the Gibbard-Satterthwaite theorem. Economic Letters* **70** (2001), 381–5.
Shepsle, K. and M. Bonchek, *Analyzing politics: rationality, behavior, and institutions*. Norton: New York and London, 1997.
Smith, D. *Manipulability measures of common social choice functions. Social Choice and Welfare* **16** (1999), 639–61.
Straffin, P. *Topics in the theory of voting*. Birkhauser: Boston, 1980.
Tanaka, Y. *An alternative direct proof of Gibbard's random dictatorship theorem*, preprint, 2004.
Taylor, A. *Mathematics and politics: strategy, voting, power, and proof.* Springer-Verlag: New York, 1995.
Taylor, A. 2002. *The manipulability of voting systems. The American Mathematical Monthly* **109**, 321–37.
Taylor, A. and W. Zwicker, *Simple games: desirability relations, trading, and pseudoweightings*. Princeton University Press: Princeton, NJ, 1999.
von Neumann, J. *Zur Theorie der Gesellschaftsspiele. Mathematische Annalen* **100** (1928), 295–320.
von Neumann, J. and O. Morgenstern, *Theory of games and economic behavior*. Princeton University Press: Princeton, NJ, 1944.
Wilson, R. *Social choice theory without the Pareto principle. Journal of Economic Theory* **3** (1972), 478–86.
Young H. P. *Social choice scoring functions. SIAM Journal on Applied Mathematics* **28** (1975), 824–38.
Young, H. P. and A. Levenglick. *A consistent extension of Condorcet's election principle. SIAM Journal on Applied Mathematics* **35** (1978), 285–300.

Index